Natti S. Rao

Diagnostics of Extrusion Processes

HANSER

Hanser Publications, Cincinnati Hanser Publishers, Munich

Distributed in North and South America by:
Hanser Publications
6915 Valley Avenue, Cincinnati, Ohio 45244-3029, USA
Fax: (513) 527-8801
Phone: (513) 527-8977
www.hanserpublications.com

Distributed in all other countries by
Carl Hanser Verlag
Postfach 86 04 20, 81631 Munich, Germany
Fax: +49 (89) 98 48 09
www.hanser-fachbuch.de

The use of general descriptive names, trademarks, etc., in this publication, even if the former are not especially identified, is not to be taken as a sign that such names, as understood by the Trade Marks and Merchandise Marks Act, may accordingly be used freely by anyone. While the advice and information in this book are believed to be true and accurate at the date of going to press, neither the author nor the editors nor the publisher can accept any legal responsibility for any errors or omissions that may be made. The publisher makes no warranty, express or implied, with respect to the material contained herein.

Cataloging-in-Publication Data is on file with the Library of Congress

ISBN 978-1-56990-568-5
E-Book-ISBN 978-1-56990-569-2

Bibliografische Information Der Deutschen Bibliothek
Die Deutsche Bibliothek verzeichnet diese Publikation in der Deutschen Nationalbibliografie; detaillierte bibliografische Daten sind im Internet über <http://dnb.d-nb.de> abrufbar.

All rights reserved. No part of this book may be reproduced or transmitted in any form or by any means, electronic or mechanical, including photocopying or by any information storage and retrieval system, without permission in writing from the publisher.

© Carl Hanser Verlag, Munich 2014
Production Management: Steffen Jörg
Coverdesign: Stephan Rönigk
Typesetted, printed and bound by Kösel, Krugzell
Printed in Germany

Table of Contents

Preface 7

1   Rheological Properties of Molten Polymers .. 11

1.1   Polymer Melt Flow 11
    1.1.1   Apparent Shear Rate 12
    1.1.2   Apparent Viscosity 13
    1.1.3   Power Law of Ostwald and de Waele 14
    1.1.4   Viscosity Formula of Klein 16
    1.1.5   Resin Characterization by Power Law Exponent ... 17
    1.1.6   Melt Flow Index 19

1.2   Relationship between Flow Rate and Pressure Drop 20

1.3   Shear Rates for Extrusion Dies 22

2   Analytical Procedures for Troubleshooting Extrusion Screws 27

2.1   Three-Zone Screw 28
    2.1.1   Extruder Output 29
    2.1.2   Feed Zone 30
    2.1.3   Metering Zone (Melt Zone) 32
    2.1.4   Practical Evaluation of Screw Geometry 33

2.2   Melting of Solids as a Tool for Solving Screw Problems 39
    2.2.1   Obtaining Better Melt Quality 39

3 Investigating Die Performance and Die Design by Computational Tools 49

3.1 Spider Dies 49
3.1.1 Pressure Drop along the Spider 50

3.2 Spiral Dies 53
3.2.1 Problem Solving by Simulating Shear Rate and Pressure 53

3.3 Adapting Die Design to Avoid Melt Fracture 56
3.3.1 Pelletizer Dies 56
3.3.2 Blow Molding Dies 56
3.3.3 Summary of the Die Design Procedures 58

3.4 Flat Dies 59

3.5 An Easily Applicable Method of Designing Screen Packs for Extruders 63
3.5.1 Mesh Size 63
3.5.2 Design Procedure 65
3.5.3 Influence of Polymer Type and Screen Blocking .. 67

4 Parametrical Studies 73

4.1 Blown Film 73

5 Design Software 79

5.1 Input and Output Data 80
5.1.1 VISRHEO 81
5.1.2 TEMPMELT 83

6 Thermal Properties of Solid and Molten Polymers ... 89

- 6.1 Specific Volume ... 89
- 6.2 Specific Heat ... 93
- 6.3 Thermal Expansion Coefficient ... 95
- 6.4 Enthalpy ... 97
- 6.5 Thermal Conductivity ... 98
- 6.6 Thermal Diffusivity ... 100
- 6.7 Coefficient of Heat Penetration ... 103
- 6.8 Heat Deflection Temperature ... 104
- 6.9 Vicat Softening Point ... 106

7 Heat Transfer in Plastics Processing ... 109

- 7.1 Case Study: Analyzing Air Gap Dynamics in Extrusion Coating by Means of Dimensional Analysis ... 109
  - 7.1.1 Heat Transfer between the Film and the Surrounding Air ... 111
  - 7.1.2 Chemical Kinetics ... 113
  - 7.1.3 Evaluation of the Experiments ... 114

References ... 117

Index ... 121

Preface

The plastics engineer working on the shop floor of an industry that manufactures blown film, blow-molded articles, or injection-molded parts, to name a few processes, often needs quick answers to questions, such as why extruder output is lower than expected, whether changing the resin will produce a better quality product, or how to estimate the pressure drop in the die. Numerical analysis to address these kinds of issues is time-consuming and costly, and requires trained personnel. As experience shows, most of these issues can be addressed quickly by applying proven, practical calculation procedures that can be performed by pocket calculators, and therefore can be handled on the site where the machines are running.

This book is an abridged version of the same topic treated in *Understanding Plastics Engineering Calculations* by Natti S. Rao and Nick R. Schott. Yet in this compact format it has all the features of the original work and can be used, not only for estimating the effect of design and process parameters on the product quality, but also for troubleshooting practical problems encountered in the field of polymer processing by extrusion. This book is intended for beginners as well as for practicing engineers, students, and teachers in the field of plastics engineering, and also for scientists from other areas who deal with polymer engineering in their professions.

The underlying principles of design formulas for plastics engineers, with examples, have been treated in detail in the earlier works of Natti Rao.

Bridging the gap between theory and practice, this book presents analytical methods based on these formulas that enable the plastics engineer to solve everyday problems related to machine design and process optimization quickly. The diagnostic approach used here is in examining whether the machine design and the resin properties are suited to each other to achieve the desired process targets.

In order to facilitate easy use, the formulas have been repeated in some calculations so that the reader does not need to refer back to the same formulas given elsewhere in the book.

Chapter 1 deals with the rheological properties of polymer melts and their use in designing polymer machinery, illustrating the topic with a number of easily understandable practical examples.

In Chapter 2, starting from melting in the screw, the relationship between screw design and the melt homogeneity is first explained by means of problems in processing taken from blown film, thermoforming, and extrusion coating. It is then shown how product quality can be improved by redesigning the screw using analytical software based on extrusion modeling techniques. Design and performance of spider dies, spiral dies, and flat film dies are the topics of Chapter 3. Here again, software based on analytical procedures evaluates the performance of the die and optimizes the design in order to meet the target values such as desired die pressure. Examples given relate to blown film and blow molding. Parametrical studies in practice on the shop floor are an important tool for optimizing the process.

Taking blown film as an example, the effect of various parameters such as film cooling and blow up are illustrated in Chapter 4. The features of two software programs that can be applied to design and optimize extrusion machinery are explained in Chapter 5. The resin data bank that is used for storing polymer property data is described in detail. Chapter 6 covers the thermodynamic properties of polymers, necessary for doing heat transfer calculations when designing polymer machinery.

Chapter 7 is a case study on the application of heat transfer and dimensionless analysis for predicting the relative importance of the parameters, such as melt temperature and dwell time of melt in the air gap in an extrusion coating process.

Thanks are due to Dr. Benjamin Dietrich of The Karlsruhe Institute of Technology for his cooperation in preparing the manuscript, and to Professor Nick R. Schott, Emeritus Professor at the University of Massachusetts at Lowell, for fruitful suggestions.

Natti S. Rao, Ph.D.

1 Rheological Properties of Molten Polymers

The basic principle of making parts from polymeric materials is in creating a melt from the solid material and forcing the melt into a die, the shape of which corresponds to the shape of the part. Thus, as Fig. 1.1 indicates, melt flow and heat transfer play an important role in the operations of polymer processing.

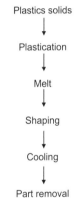

Figure 1.1 Principle of manufacturing of plastics parts

1.1 Polymer Melt Flow

Macromolecular fluids, such as thermoplastic melts, exhibit significant non-Newtonian behavior. This is noticed in the marked decrease of melt viscosity when the melt is subjected to shear or tension as shown in Figs. 1.2 and 1.4. Melt flow in the channels of dies and polymer processing machinery is mainly shear flow. Therefore, knowledge of the laws of shear flow is necessary for designing machines and dies for polymer processing. For practical applications, the following summary of the relationships was found to be useful.

1.1.1 Apparent Shear Rate

The apparent shear rate for a melt flowing through a capillary is defined as

$$\dot{\gamma}_a = \frac{4\dot{Q}}{\pi R^3} \tag{1.1}$$

where $\dot{Q}$ is the volumetric flow rate per second and R is the radius of the capillary. This relationship is for *steady state*, incompressible flow without entrance or exit effects, no wall slip, and with symmetry about the center line.

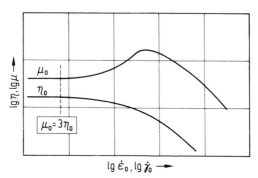

Figure 1.2 Tensile viscosity and shear viscosity of a polymer melt as a function of strain rate [1]

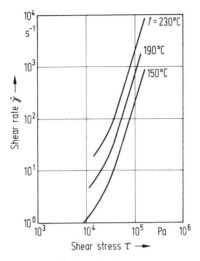

Figure 1.3 Flow curves of a LDPE [2]

1.1.2 Apparent Viscosity

The apparent viscosity η_a is defined as

$$\eta_a = \frac{\tau}{\dot{\gamma}_a} \tag{1.2}$$

and is shown in Fig. 1.4 as a function of shear rate and temperature for a LDPE.

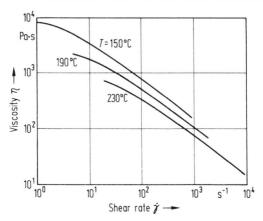

Figure 1.4 Viscosity functions of a LDPE [3]

1.1.3 Power Law of Ostwald and de Waele

The power law of Ostwald [4] and de Waele [5] is easy to use; hence, it is widely employed in design work [2]. This relationship can be expressed as

$$\dot{\gamma}_a = K\tau^n \tag{1.3}$$

or

$$\dot{\gamma}_a = K\left|\tau^{n-1}\right|\tau \tag{1.4}$$

where K denotes a factor of proportionality and n the power law exponent. Another form of the power law that is often used is

$$\tau_a = K_R\, \dot{\gamma}_a^{n_R} \tag{1.5}$$

or $\tau_a = K_R \left|\dot{\gamma}_a^{n_R-1}\right|\dot{\gamma}_a$ (1.6)

In this case, n_R is the reciprocal of n and $K_R = K^{-n_R}$. In the US, n_R is used instead of n.

From Eq. 1.3 the exponent n can be expressed as

$$n = \frac{d \lg \dot{\gamma}_a}{d \lg \tau} \quad (1.7)$$

where lg means logarithm to the base of 10, which is the same throughout this book.

As shown in Fig. 1.5, in a double log-plot, the exponent n represents the local gradient of the curve $\dot{\gamma}_a$ vs. τ.

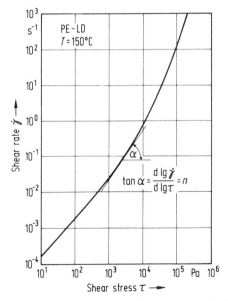

Figure 1.5 Determination of the power law exponent n in the Eq. 1.7

1.1.4 Viscosity Formula of Klein

Due to the ease of its application, the Klein model, among other rheological models, is best suited for practical use.

The regression equation of Klein et al. [6] is given by

$$\lg \eta_a = a_0 + a_1 \ln \dot{\gamma}_a + a_{11}(\ln \dot{\gamma}_a)^2 + a_2 T + a_{22} T^2 + a_{12} T \ln \dot{\gamma}_a \tag{1.8}$$

$T =$ Temperature of the melt (°F)
$\eta_a =$ Viscosity (lb$_f \cdot$ s/in^2)

ln means natural logarithm throughout the book.

The resin-dependent constants a_0 to a_{22} can be determined with the help of the computer program VISRHEO discussed in Chapter 5.

Calculated Example

The following constants are valid for a particular type of LDPE. What is the viscosity η_a at $\dot{\gamma}_a = 500$ s^{-1} and $T = 200$ °C?

a_0	=	3.388
a_1	=	$-6.351 \cdot 10^{-1}$
a_{11}	=	$-1.815 \cdot 10^{-2}$
a_2	=	$-5.975 \cdot 10^{-3}$
a_{22}	=	$-2.51 \cdot 10^{-6}$
a_{12}	=	$5.187 \cdot 10^{-4}$

T (°F) $= 1.8 \cdot T$ (°C) $+ 32 = 1.8 \cdot 200 + 32 = 392$

With the above constants and Eq. 1.8, one gets

$\eta_a = 0.066$ lb$_f \cdot$ s/in^2

and in SI-units,

$\eta_a = 6857 \cdot 0.066 = 449.8$ Pa · s

The expression for the power law exponent n can be derived from Eq. 1.8. The exponent n is given by

$$\frac{1}{n} = 1 + a_1 + 2a_{11} \ln \dot{\gamma}_a + a_{12} \cdot T \tag{1.9}$$

Putting the constants $a_1, ..., a_{12}$ into this equation one obtains the power law exponent

$n = 2.919$

1.1.5 Resin Characterization by Power Law Exponent

It is seen from Fig. 1.6 that n depends less and less on the shear rate with increasing shear rate, so that using a single value of n for a given resin leads to a good estimate of the parameter desired. Plots of n for LLDPE and PET are given in Figs. 1.7 and 1.8.

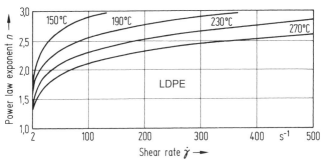

Figure 1.6 Power law exponent n as a function of shear rate $\dot{\gamma}$ and melt temperature T [2]

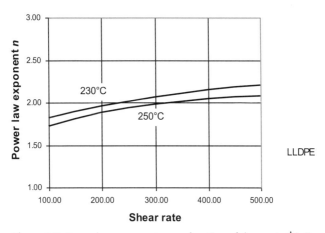

Figure 1.7 Power law exponent n as a function of shear rate $\dot{\gamma}$ (s^{-1}) and temperature for LLDPE

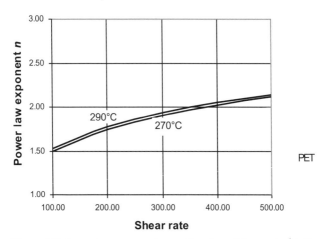

Figure 1.8 Power law exponent n as a function of shear rate $\dot{\gamma}$ (s^{-1}) and temperature for PET

1.1.6 Melt Flow Index

The *Melt Flow Index* (MFI), which is also known as the *Melt Flow Rate* (MFR), indicates the flowability of a constant polymer melt and is measured by forcing the melt through a capillary under a dead load at constant temperature (Fig. 1.9). The MFI value is the mass of melt flowing in a certain time. A MFR or MFI of 2 at 200 °C and 2.16 kg means, for example, that the melt at 200 °C flows at a rate of 2 g in ten minutes under a dead load of 2.16 kg.

In the case of melt volume rate, which is also known as *Melt Volume Index* (MVI), the volume flow rate of the melt instead of the mass flow rate is set as the basis. The unit here is ml/10 min.

Ranges of melt indices for common processing operations are given in Table 1.1 [7].

Table 1.1 Ranges of MFI Values (ASTM D1238) for Common Processes [7]

Process	MFI range
Injection molding	5–100
Rotational molding	5–20
Film extrusion	0.5–6
Blow molding	0.1–1
Profile extrusion	0.1–1

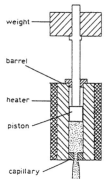

Figure 1.9 Melt flow tester [8]

1.2 Relationship between Flow Rate and Pressure Drop

To illustrate the behavior of polymer melt flow, the flow rate is plotted against the die pressure for water and LDPE in Fig. 1.10. The diameter and length of the nozzle used are

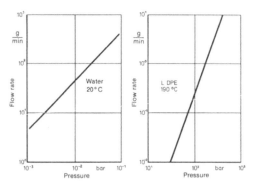

Figure 1.10 Flow curves of water in laminar flow and LDPE [9]

1 mm and 30 mm, respectively. Figure 1.10 shows that the laminar flow rate of water increases linearly with the pressure, whereas in the case of LDPE, the increase is exponential. To put it in numbers, increasing the pressure of the water tenfold would bring forth a tenfold flow rate. A tenfold pressure increase of LDPE, however, leads to about a five-hundredfold flow rate. The flow of the melt is indeed viscous, but pressure changes are accompanied by much larger changes in output. The relationship between volume flow rate and pressure drop of the melt in a die can be expressed in the general form [10]

$$\dot{Q} = K \cdot G^n \cdot \Delta p^n z \tag{1.10}$$

where Q = volumetric flow rate
 G = die constant
 Δp = pressure difference
 K = factor of proportionality
 n = power law exponent

The approximate ranges of shear rates for different methods of polymer processing are shown in Fig. 1.11 [7].

Choosing a value of $n = 2.7$ for LDPE, Eq. 1.10 can be written for a given die and processing conditions as

$$\dot{Q} \sim \Delta p^{2.7} \tag{1.11}$$

Increasing the pressure by tenfold would increase the volume flow rate by about five-hundredfold ($10^{2.7} = 501.2$).

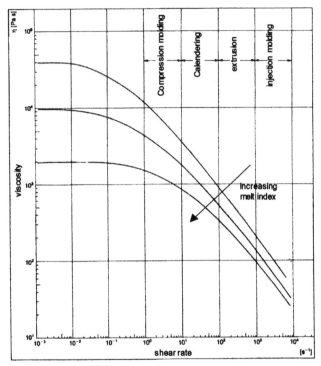

Figure 1.11 Ranges of shear rates for different methods of processing for LDPE [7]

1.3 Shear Rates for Extrusion Dies

The formulas for calculating the shear rates in extrusion dies are presented in Table 1.2 [2, 10] and the channel shapes in Fig. 1.12 [11]. Other channel shapes can be taken into account by the formula developed by Schenkel [12].

Table 1.2 Shear Rates and Geometry Constants for Some Die Channel Shapes

Channel Shape	Shear rate $\dot{\gamma}$ [s^{-1}]	Geometry Constant G
Circle	$4\dot{Q}/\pi R^3$	$\left(\dfrac{\pi}{4}\right)^{\frac{1}{n}} \cdot \dfrac{R^{\frac{3}{n}+1}}{2L}$
Slit	$\dfrac{6\dot{Q}}{W \cdot H^2}$	$\left(\dfrac{W}{6}\right)^{\frac{1}{n}} \cdot \dfrac{H^{\frac{2}{n}+1}}{2L}$
Annulus	$\dfrac{6\dot{Q}}{\pi (R_o + R_i)(R_o - R_i)^2}$	$\left(\dfrac{\pi}{6}\right)^{\frac{1}{n}} \cdot \dfrac{(R_o + R_i)^{\frac{1}{n}} \cdot (R_o - R_i)^{\frac{2}{n}+1}}{2L}$
Triangle	$\dfrac{10}{3} \cdot \dfrac{\dot{Q}}{d^3}$	$\dfrac{1}{\sqrt{3}} \cdot \left(\dfrac{3}{10}\right)^{\frac{1}{n}} \cdot \dfrac{d^{\frac{3}{n}+1}}{2L}$
Square	$\dfrac{3}{0.42} \cdot \dfrac{\dot{Q}}{a^3}$	$\dfrac{1}{2}\left(\dfrac{0.42}{3}\right)^{\frac{1}{n}} \cdot \dfrac{a^{\frac{3}{n}+1}}{2L}$

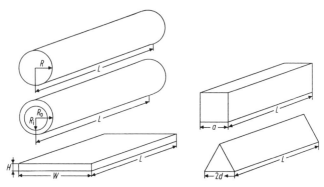

Figure 1.12 Common shapes of flow channels in extrusion dies [11]

The following examples illustrate the use of the formulas given above.

Calculated Examples

Example 1
What is the shear rate of a LDPE melt at 200 °C flowing through a round channel of 25 mm diameter at a mass flow rate of $\dot{m} = 36$ kg/h with a melt density of $\rho_n = 0.7$ g/cm³?

Solution
Volume shear rate

$$\dot{Q} = \dot{m}/\rho_n = 1.429 \; 10^{-5} \; m^3/s$$

Shear rate

$$\dot{\gamma} = 4\dot{Q}/(\pi)R^3 = 9.361 \; s^{-1}$$

Example 2

Melt flow through an annulus of a blown film die with an outside radius $R_o = 40$ mm and an inside radius $R_i = 39$ mm. The resin is LDPE with the same viscosity as above. Mass flow rate and the melt temperature remain the same.

Solution
Volume shear rate

$$\dot{Q} = \dot{m}/\rho_n = 1.429 \; 10^{-5} \; m^3/s$$

Shear rate

$$\dot{\gamma} = 6\dot{Q}/\left(\pi(R_o + R_i)(R_o - R_i)^2\right) = 345.47 \; s^{-1}$$

Example 3

With the same conditions as above, it is required to calculate the shear rate when the melt flows through a slit of width $W = 75$ mm and height $H = 1$ mm of an extrusion coating die.

Solution
Shear rate

$$\dot{\gamma} = 6\dot{Q}/\left(W H^2\right) = 1143.2 \; s^{-1}$$

The calculated shear rates in different dies enable one to obtain the shear viscosity from the plot η vs. $\dot{\gamma}$, which then can be used in a comparative study of the resin behavior.

Example 4

Calculate the pressure drop, Δp, for the conditions given in Example 2. Using the Klein model [6]

$$\ln \eta = a_0 + a_1 \ln \dot\gamma + a_{11} \ln \dot\gamma^2 + a_2 T + a_{22} T^2 + a_{12} T \ln \dot\gamma$$

with viscosity coefficients [6]

$a_0 = 3.388$, $a_1 = -0.635$, $a_{11} = -0.01815$,
$a_2 = -0.005975$, $a_{22} = -0.0000025$, $a_{12} = 0.0005187$

η at $\dot\gamma = 9.316$ s^{-1} is found to be $\eta = 4624.5$ Pa · s with $T = 392\,°F$. The power law exponent n follows from Eq. 1.9

$n = 2.052$

Shear stress

$\tau = \eta \cdot \dot\gamma = 43077$ N/m²

Factor of proportionality K in Eq. 1.10 from Eq. 1.3

$K = 28824 \cdot 10^{-9}$

Die constant from Table 1.2

$$G_{circle} = \left(\frac{\pi}{4}\right)^{\frac{1}{2.052}} \frac{0.0125^{\frac{3}{2.052}+1}}{2 \cdot 0.1} = 9.17 \cdot 10^{-5}$$

Finally Δp from Eq. 1.10

$$\Delta p = \frac{\left(1.429 \cdot 10^{-5}\right)^{\frac{1}{2.052}}}{\left(2.8824 \cdot 10^{-9}\right)^{\frac{1}{2.052}} \cdot 9.17 \cdot 10^{-5}} = 6.845 \text{ bar}$$

2 Analytical Procedures for Troubleshooting Extrusion Screws

Extrusion is one of the most widely used polymer converting operations for manufacturing blown film, pipes, sheets, and laminations, to name the most significant industrial applications. Figure 2.1 shows a modern large scale machine for making blown film. The extruder, which constitutes the central unit of these machines, is shown in Fig. 2.2. The polymer is fed into the hopper in the form of granulate or powder. It is kept at the desired temperature and humidity by controlled air circulation. The solids are conveyed by the rotating screw and slowly melted, in part by barrel heating, but mainly by the frictional heat generated by the shear between the polymer and the barrel (Fig. 2.3). The melt at the desired temperature and pressure flows through the die, in which the shaping of the melt into the desired shape takes place.

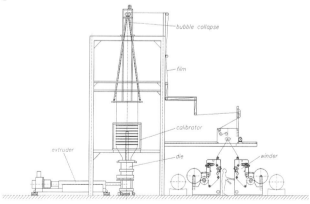

Figure 2.1 Large scale blown film line [13]

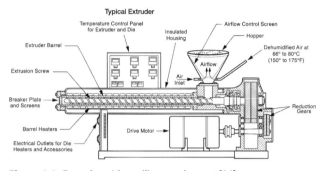

Figure 2.2 Extruder with auxiliary equipment [14]

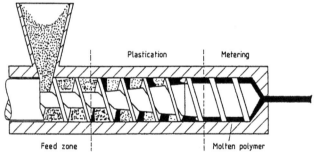

Figure 2.3 Plasticating extrusion [15]

2.1 Three-Zone Screw

Basically extrusion consists of transporting the solid polymer in an extruder by means of a rotating screw, melting the solid, homogenizing the melt, and forcing the melt through a die (Fig. 2.3).

The extruder screw of a conventional plasticating extruder

has three geometrically different zones (Fig. 2.4), with functions that can be described as follows:

Feed zone: transport and preheating of the solid material

Transition zone: compression and plastication of the polymer

Metering zone: melt conveying, melt mixing, and pumping of the melt to the die

However, the functions of a zone are not limited to that particular zone alone. The processes mentioned can continue to occur in the adjoining zone as well.

Although the following equations apply to the three-zone screws, they can be used segment-wise for designing screws of other geometries as well.

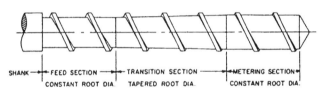

Figure 2.4 Three-zone screw [8]

2.1.1 Extruder Output

Depending on the type of extruder, the output is determined either by the geometry of the solids feeding zone alone, as in the case of a grooved extruder [16], or by the solids and melt zones to be found in a smooth barrel extruder. A too high or too low output results when the dimensions of the screw and die are not matched with each other.

2.1.2 Feed Zone

A good estimate of the solids flow rate can be obtained from Eq. 2.1 as a function of the conveying efficiency and the feed depth. The desired output can be found by simulating the effect of these factors on the flow rate by means of Eq. 2.1.

Calculated Example

The solids transport is largely influenced by the frictional forces between the solid polymer and barrel and screw surfaces. A detailed analysis of the solids conveying mechanism was performed by Darnell and Mol [17]. The following example presents an empirical equation that provides good results in practice [2].

The geometry of the feed zone of a screw (Fig. 2.5) is given by the following data:

Barrel diameter	D_b = 30 mm
Screw lead	s = 30 mm
Number of flights	ν = 1
Flight width	w_{FLT} = 3 mm
Channel width	W = 28.6 mm
Depth of the feed zone	H = 5 mm
Conveying efficiency	η_F = 0.436
Screw speed	N = 250 rpm
Bulk density of the polymer	ρ_o = 800 kg/m^3

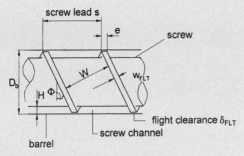

Figure 2.5 Screw zone of a single screw extruder [11]

The solids conveying rate in the feed zone of the extruder can be calculated according to [15]

$$G = 60 \cdot \rho_o \cdot N \cdot \eta_F \cdot \pi^2 \cdot H \cdot D_b (D_b - H) \frac{W}{W + w_{FLT}} \quad (2.1)$$
$$\cdot \sin\phi \cdot \cos\phi$$

with the helix angle ϕ

$$\phi = \tan^{-1}[s/(\pi \cdot D_b)] \quad (2.2)$$

The conveying efficiency η_F in Eq. 2.1 as defined here is the ratio between the actual extrusion rate and the theoretical maximum extrusion rate attainable under the assumption of no friction between the solid polymer and the screw. It depends on the type of polymer, bulk density, barrel temperature, and the friction between the polymer, barrel, and the screw. Experimental values of η_F for some polymers are given in Table 2.1.

Table 2.1 Conveying Efficiency η_F for Some Polymers

Polymer	Smooth Barrel	Grooved Barrel
LDPE	0.44	0.8
HDPE	0.35	0.75
PP	0.25	0.6
PVC-P	0.45	0.8
PA	0.2	0.5
PET	0.17	0.52
PC	0.18	0.51
PS	0.22	0.65

Using the values above, with the dimensions in meters, in Eqs. 2.1 and 2.2 we get

$$G = 60 \cdot 800 \cdot 250 \cdot 0.44 \cdot \pi^2 \cdot 0.005 \cdot 0.03 \cdot 0.025$$

$$\cdot \frac{0.0256}{0.0286} \cdot 0.3034 \cdot 0.953$$

Hence $G \approx 50$ kg/h

2.1.3 Metering Zone (Melt Zone)

The conveying capacity of the melt zone must correspond to the amount of melt created by plastication, in order to ensure stable melt flow without surging. This quantity can be estimated by means of the equations presented in the book under further reading.

2.1.4 Practical Evaluation of Screw Geometry

Based on the laws of similarity, Pearson [18] developed a set of relationships to scale-up a single screw extruder. These relationships are useful for the practicing engineer to estimate the size of a larger extruder from experimental data gathered on a smaller machine. The scale-up assumes equal length to diameter ratios between the two extruders. The important relations can be summarized as follows:

$$\frac{H_2}{H_1} = \left(\frac{D_2}{D_1}\right)^{(1-s)/(2-3s)} \tag{2.3}$$

$$\frac{N_2}{N_1} = \left(\frac{D_2}{D_1}\right)^{-(2-2s)/(2-3s)} \tag{2.4}$$

$$\frac{\dot{m}_2}{\dot{m}_1} = \left(\frac{D_2}{D_1}\right)^{(3-5s)/(2-3s)} \tag{2.5}$$

$$\frac{H_{F_2}}{H_{F_1}} = \left(\frac{D_2}{D_1}\right)^{(1-s)/(2-3s)} \tag{2.6}$$

Where H_F = feed depth
H = metering depth
D = screw diameter
N = screw speed

Subscripts 1 and 2 equal a screw of known geometry and a screw for which the geometry is to be determined, respectively.

The exponent s is given by

$$s = 0.5(1 - n_R) \tag{2.7}$$

where n_R is the reciprocal of the power law exponent n. The shear rate required to determine n is obtained from

$$\dot{\gamma}_a = \frac{\pi \cdot D_1 \cdot N_1}{60 \cdot H_1} \tag{2.8}$$

Calculated Example

The following conditions are given:
The resin is LDPE and the stock temperature is 200 °C. The data pertaining to Screw 1 are:

$D_1 = 90$ mm; $H_F = 12$ mm; $H_1 = 4$ mm
Feed length   = $9\,D_1$
Transition length = $2\,D_1$
Metering length = $9\,D_1$
Output $\dot{m}_1$ = 130 kg/h
Screw speed N_1 = 80 rpm

The diameter of Screw 2 is $D_2 = 120$ mm. The geometry of Screw 2 is to be determined.

Solution
The geometry is computed from the equations given above [2]. It follows that

$D_2 = 120$ mm
$H_{F_2} = 14.41$ mm

$H_2 = 4.8$ mm
$\dot{m}_2 = 192.5$ kg/h
$N_1 = 55.5$ rpm

Other methods of scaling up have been treated by Schenkel [12], Fischer [19], and Potente [20].

Examples for calculating the dimensions of extrusion screws are given below.

Calculated Example

Determine the specific output for a LLDPE resin with a power law exponent, $n = 2$.

Solution
Substituting $n = 2$ into Eq. 2.7, we get $s = 0.25$. Using Eqs. 2.4 and 2.5, we finally obtain

$$\dot{m}_s = \frac{\dot{m}}{N} = 0.0000155 \, D^{2.6} \tag{2.9}$$

where $\dot{m}$ = output kg/h
N = screw speed rpm
D = screw diameter mm
$\dot{m}_s$ = specific output

The specific output, $\dot{m}_s$, is plotted as a function of the screw diameter D for LLDPE and LDPE in Fig. 2.6 by means of Eq. 2.9 and the formula derived for LDPE with the exponent $n = 2.5$ from Eqs. 2.4, 2.5, and 2.7.

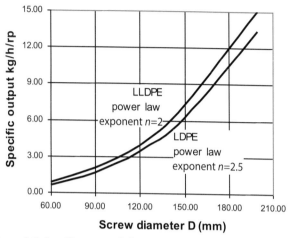

Figure 2.6 Specific output, $\dot{m}_s$ for LDPE and LLDPE as a function of screw diameter for an extruder of $L/D = 20$

Other target values of practical interest are plotted in Figs. 2.7 to 2.9 as a function of screw diameter. Machine manufacturers confirmed the application of these easy-to-use relationships, although only one single value of the power law exponent is used to characterize the resin flow.

The scale-up of extrusion screws can be performed quickly by using the program VISSCALE given in Chapter 5.

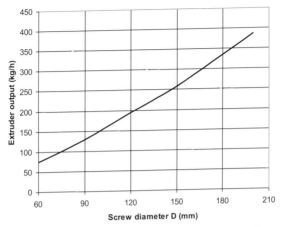

Figure 2.7 Extruder output vs. screw diameter for LDPE and extruder of $L/D = 20$, $\dot{m} = 0.281 \cdot D^{1.364}$ kg/h

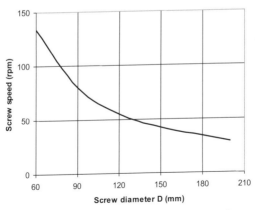

Figure 2.8 Screw speed vs. screw diameter for LDPE and extruder of $L/D = 20$, $N = 24600$, and $D^{-1.273}$ rpm

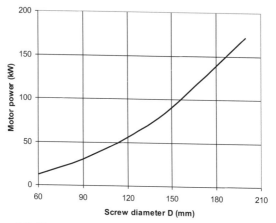

Figure 2.9 Motor power vs. screw diameter for LDPE and extruder of $L/D = 20$, PWR = $0.0015\ D^{2.2}$ kW

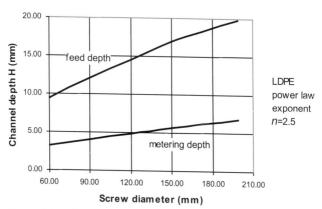

LDPE power law exponent $n=2.5$

Figure 2.10 Channel depths of the screw for LDPE for an extruder of $L/D = 20$

2.2 Melting of Solids as a Tool for Solving Screw Problems

Physical models describing the melting of solids in extruder channels were developed by many workers, notable among them, the work of Tadmor [15]. Rauwendaal summarizes the theories underlying these models in his book [8].

2.2.1 Obtaining Better Melt Quality

The performance of an extrusion screw employed in a process can be simulated by using the equations for the melting profile, stock temperature, and melt pressure given in [2], and if necessary, the screw design optimized for better performance by using the program TEMPMELT mentioned in Chapter 5. The results of a simulation are shown schematically in Fig. 2.11 [21]. The following practical examples illustrate this procedure.

Example 1: Poor Melt Quality in an Extrusion Coating Process

Figure 2.3 shows the melting of a polymer in the screw channels of an extrusion screw. The melt quality can be defined by the ratio G_s/G, as depicted in Fig. 2.11. The melting is completed, if this ratio is zero [2].

This example refers to an extrusion problem [22] concerning melt quality, which a processor had while extruding LDPE in a coating process. Computer simulations of the screw used by the processor at two different output rates showed that the solids content of the polymer at the entrance of the mixing section, particularly at the higher output, was too high to be melted and homogenized by the mixing device as shown in Fig. 2.12.

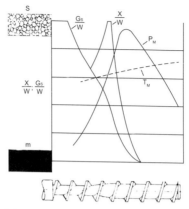

Figure 2.11 Schematic screw simulation profiles. G_s: solids flow rate, G: Total flow rate, X: Width of solids bed W: Channel width, P_M: Melt Pressure, T_M: Melt temperature [21]

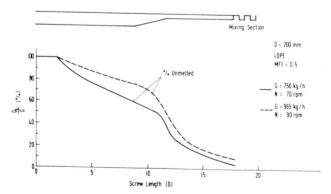

Figure 2.12 Poor melt quality of a processor's screw

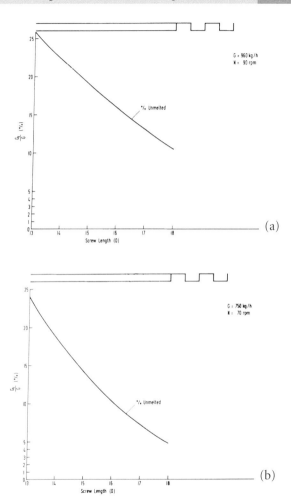

Figure 2.13 Enlarged melting profiles of the customer's screw at low (a) and high (b) outputs

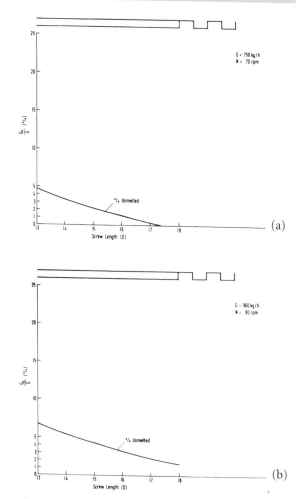

Figure 2.14 Enlarged melting profiles of the redesigned screw at low (a) and high (b) outputs

The enlarged melting profiles in the metering zone presented in Fig. 2.14 clearly show that inhomogeneous melt results from the improper screw design and that the screw employed by the processor cannot produce a melt of good quality at the higher output desired by the processor.

The screw was redesigned by implementing different screw geometries with a mixing device (Fig. 2.16) into the program. As the enlarged melting profiles in the metering zone of this screw (given in the Fig. 2.14) show, the solids content at the inlet of the mixing device is now low enough to produce a homogeneous melt.

The processor obtained acceptable melt quality at the desired output by using the redesigned screw with shearing and distributive mixers, confirming the predictions of the program, as shown in Fig. 2.15.

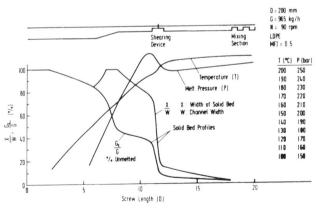

Figure 2.15 Redesigned screw with better melt homogeneity

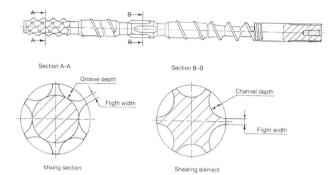

Figure 2.16 Screw configuration with shearing and mixing devices

Example 2: Determination of the Optimal Screw Geometry from a Given Number of Screws of Different Geometry Used in a Blown Film Process

The program TEMPMELT can also be used to select a screw best suited for given operating conditions, among a set of existing screws of varied geometry.

A processor had five three-zone standard extrusion screws of different geometries at his disposal, as shown in Fig. 2.17. The aim was to determine the screw geometry best suited for the operating conditions mentioned in Fig. 2.17. The calculated melting profiles in Fig. 2.17 show that Screw 5 has the optimal geometry, because complete melting has been attained. This predicted result was confirmed in practice.

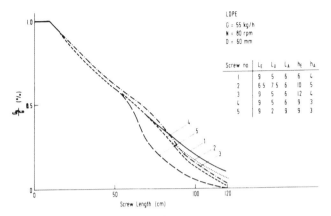

Figure 2.17 Determining optimal screw geometry

Example 3: Improving the Quality of Melt in an Extruder Used for Extrusion Coating and Laminating

This example deals with a screw used by a customer in a coating line for processing PET resin. It was found that the quality of the melt created by the screw was poor. As shown in Fig. 2.18, the simulation confirmed this observation of the customer.

The computer program TEMPMELT also predicted that the quality of the melt can be improved by increasing the temperature of the pellets from 20 to 80 °C.

This measure was then employed in practice and led to a melt of better quality to the satisfaction of the customer, see Fig. 2.18(b).

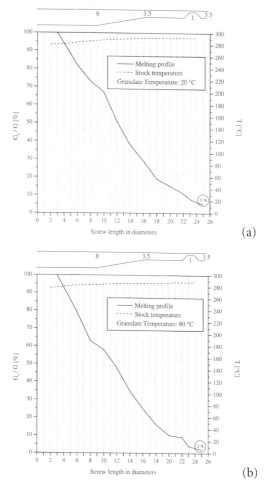

Figure 2.18 Melt quality at low (a) and high (b) granulate temperature

Example 4: Air Bubbles in a Flat Film Used in Packaging

Figure 2.19 shows air bubbles in a flat film of PA-6 exiting from a coat hanger die. By using different barrel temperature profiles, it was found that the temperature profile given in Fig. 2.20 with a high temperature near the feed zone would eliminate the bubbles. The predicted result was confirmed in practice.

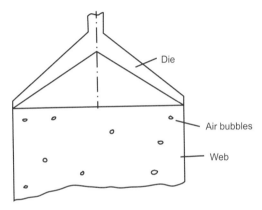

Figure 2.19 Air bubbles in a flat film of PA-6

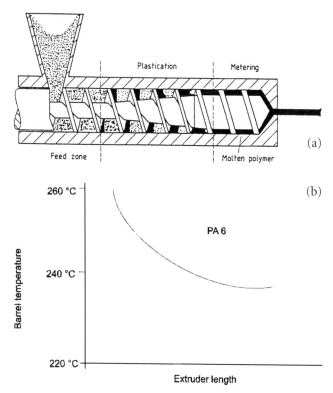

Figure 2.20 Barrel temperature used to eliminate air bubbles

The examples presented here show that the use of computer programs in design work minimizes experimentation and in many cases eliminates it altogether.

3 Investigating Die Performance and Die Design by Computational Tools

The design of extrusion dies is based on the principles of rheology, thermodynamics, and heat transfer as described in the book by Natti Rao and Nick Schott, which is suggested for further reading, and listed along with the references at the end of this book. The quantities to be calculated are pressure, shear rate, and residence time along the die length. The most important parameter among these is, however, the die pressure as this quantity must match with the pressure created by the extruder.

3.1 Spider Dies

Spider dies are used in blown film, pipe extrusion, and blow molding processes. The shear rate and the die constant for circle and annulus can be calculated from the relationships, given in Table 1.2. The pressure drop along the die

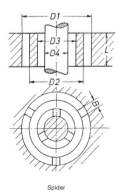

Figure 3.1 Spider cross sections

follows from Eq. 1.10, as illustrated in the Examples 2 and 4 in Chapter 1. In general, the die consists of the geometrical sections given in Fig. 3.1. By using the program VISPIDER, the performance of a spider die can be easily simulated for any polymer without having to conduct any experiments with the die.

3.1.1 Pressure Drop along the Spider

Example: Calculating Pressure Drop

For the input data shown in Table 3.1, the pressure drop along the die length for different die gaps is calculated using the equations for the round channel and annulus given in [2] by using the program VISPIDER as shown in the Fig. 3.2.

Table 3.1 Pressure Drop vs. Spider Length

```
FLOW RATE:              30.00000  (KG/H)

D1:                     45.00000  (MM)
D2:                     16.00000  (MM)
L:                     100.00000  (MM)
PR.DROP        :          9.65085 (BAR)

D1:                     16.00000  (MM)
D2:                     16.00000  (MM)
L:                      40.00000  (MM)
PR.DROP        :         11.02954 (BAR)

D1:                     16.00000  (MM)
D2:                     16.00000  (MM)
L:                      46.00000  (MM)
PR.DROP        :         12.68397 (BAR)

D1:                     16.00000  (MM)
D2:                     63.00000  (MM)
D3:                      0.0      (MM)
D4:                     28.00000  (MM)
L:                      36.00000  (MM)
PR.DROP        :          3.65644 (BAR)

D1:                     68.00000  (MM)
D2:                     54.00000  (MM)
D3:                     42.00000  (MM)
D4:                     28.00000  (MM)
L:                      33.00000  (MM)
```

Table 3.1 *(Continued):* Pressure Drop vs. Spider Length

```
N :                  3.00900  (--)
B :                  6.00000  (MM)
PR.DROP     :       13.47862  (BAR)

D1:                 68.00000  (MM)
D2:                 68.00000  (MM)
D3:                 28.00000  (MM)
D4:                 44.00000  (MM)
L :                 38.00000  (MM)
PR.DROP     :        1.48398  (BAR)

D1:                 68.00000  (MM)
D2:                 54.00000  (MM)
D3:                 44.00000  (MM)
D4:                 35.00000  (MM)
L :                 30.00000  (MM)
PR.DROP     :        2.62504  (BAR)

D1:                 54.00000  (MM)
D2:                 49.00000  (MM)
D3:                 35.00000  (MM)
D4:                 28.00000  (MM)
L :                  8.00000  (MM)
PR.DROP     :        0.87935  (BAR)

D1:                 49.00000  (MM)
D2:                 45.90599  (MM)
D3:                 28.00000  (MM)
D4:                 28.00000  (MM)
L :                  5.50000  (MM)
PR.DROP     :        0.67009  (BAR)

D1:                 45.90599  (MM)
D2:                 40.00000  (MM)
D3:                 28.00000  (MM)
D4:                 26.04500  (MM)
L :                 10.50000  (MM)
PR.DROP     :        2.01417  (BAR)

D1:                 40.00000  (MM)
D2:                 27.09999  (MM)
D3:                 26.04500  (MM)
D4:                 19.99500  (MM)
L :                 32.50000  (MM)
PR.DROP     :       16.87939  (BAR)

D1:                 27.09999  (MM)
D2:                 26.30600  (MM)
D3:                 19.99500  (MM)
D4:                 19.46599  (MM)
L :                  1.50000  (MM)
PR.DROP     :        1.48500  (BAR)
```

Table 3.1 *(Continued):* Pressure Drop vs. Spider Length

```
D1:                        26.30600 (MM)
D2:                        18.00000 (MM)
D3:                        19.46599 (MM)
D4:                        14.00000 (MM)
L:                         15.50000 (MM)
PR.DROP       :            27.94635 (BAR)

TOTAL PR.DROP     :        104.50705 (BAR)
EXIT TEMPERATURE  :        166.21030 (GRAD C)
```

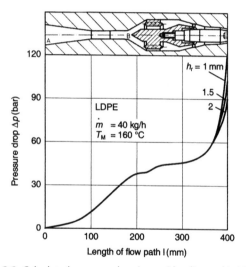

Figure 3.2 Calculated pressure drop in a spider die used in blown film at different die gaps

3.2 Spiral Dies

3.2.1 Problem Solving by Simulating Shear Rate and Pressure

Spiral mandrel dies are used in blown film processes for making films with as small a deviation in the film thickness as possible. As shown in Fig. 3.3, the polymer melt is divided into several separate feed ports which empty into spirally shaped channels of decreasing cross section cut into the mandrel. Between the die body and the land separating adjacent spirals, there is a gap, or an annulus, into which a part of the spiral flow leaks. The flow proceeds upward over the land into the next spiral. It is the mixing of annular and spiral flows that leads to uniform melt flow distribution. A further advantage of spiral mandrel dies over spider dies is that weld

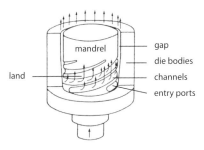

Figure 3.3 Spiral die [10]

lines and flow markings are eliminated due to the absence of mandrel support elements disturbing the flow.

A planar view of a spiral die with four channels is shown in Fig. 3.4. Here, the channels have a length which corresponds to only three quarters of the die circumference. The

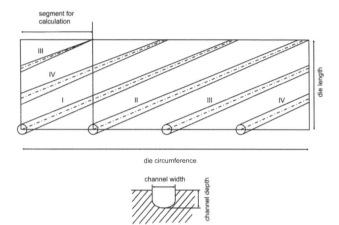

Figure 3.4 Planar view of a spiral die with four channels [23, 24]

segment of the die, which can be calculated by the computer program VISPIRAL listed in Chapter 5 using the equations given in [2], is shown in Fig. 3.4. Due to reasons of symmetry, this calculation is valid for other segments as well, so that the whole die can be designed in this manner.

Example

The input data and calculated results of a die with three channels and a spiral length of one and two thirds of the die circumference are given in Fig. 3.5 for LDPE. The die gap required for uniform melt flow, the shear rates in the spiral and annulus, as well as the pressure drop along the spiral are the output data needed to design the die.

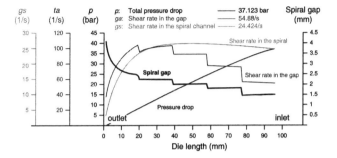

Channel pitch:	57.3 mm
Angle of pitch:	6.09°
Initial channel depth:	25 mm
Final channel depth:	6 mm
Initial channel width:	13 mm
Starting value angular gap:	1 mm
Tolerance:	1%
Increment angular gap:	.001 mm
Number of channel turns:	1.666
Number of channels:	3
Number channel elements:	100
Output:	200 kg/h
Melt temperature:	200 °C

Figure 3.5 Results of simulation of a spiral die for LDPE with input data above

3.3 Adapting Die Design to Avoid Melt Fracture

Melt fracture can be defined as an instability of the melt flow leading to surface or volume distortions of the extrudate. Surface distortions [7] are usually created from instabilities near the die exit, while volume distortions [25] originate from the vortex formation at the die entrance. Due to the occurrence of these phenomena, melt fracture limits the production of articles manufactured by extrusion processes. The use of processing additives to increase output has been written about in publications given in [26]. However, processing aids are not desirable in applications such as pelletizing and blow molding. Therefore, the effect of die geometry on the onset of melt fracture has been examined.

3.3.1 Pelletizer Dies

The aim here is to design a die for a given throughput or to calculate the maximum throughput possible without melt fracture for a given die. These targets can be achieved by performing simulations on dies of different tube diameters, flow rates, and melt temperatures. The program VISPIDER listed in Chapter 5 can be used for this case as well. Fig. 3.6 shows the results of one such simulation.

3.3.2 Blow Molding Dies

Figure 3.7 shows the surface distortion on the parison used in blow molding, which occurs at a definite shear rate depending on the resin. In order to obtain a smooth product surface, the die contour has been changed in such a way that the shear rate lies in a range that provides a smooth surface of

Adapting Die Design to Avoid Melt Fracture

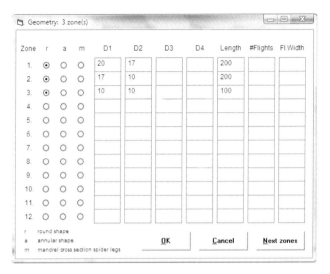

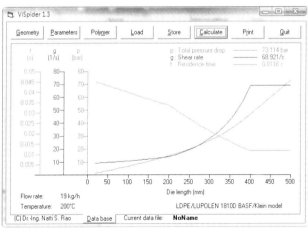

Figure 3.6 Results of simulation of a pelletizer die

Figure 3.7 Surface distortion on a parison used in the blow molding [27]

the product. The redesigned die creates lower extrusion pressures as well, as can be seen from Fig. 3.8 [27].

3.3.3 Summary of the Die Design Procedures

The following steps summarize the design procedures given.

- ▶ STEP 1: Calculation of the shear rate in the die channel
- ▶ STEP 2: Fitting the measured viscosity plots with a rheological model
- ▶ STEP 3: Calculation of the power law exponent
- ▶ STEP 4: Calculation of the shear viscosity at the shear rate in Step 1
- ▶ STEP 5: Calculation of the wall shear stress
- ▶ STEP 6: Calculation of the factor of proportionality
- ▶ STEP 7: Calculation of die constant
- ▶ STEP 8: Calculation of pressure drop in the die channel
- ▶ STEP 9: Calculation of the residence time of the melt in the channel

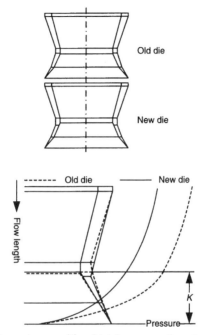

Figure 3.8 Die contour used for obtaining a smooth parison surface [27]

3.4 Flat Dies

Taking the resin behavior and the process conditions into account, the flat dies used in extrusion coating can be designed following similar guidelines as previously outlined. Figure 3.9 shows the manifold radius required to attain uniform melt flow out of the die exit as a function of the manifold length [28].

The pressure drop in the flat film die can be obtained from the relationship [2]

$$\Delta p = \frac{2\,(6\cdot\dot{Q})^{\frac{1}{n}}\cdot L}{K^{\frac{1}{n}}\cdot W^{\frac{1}{n}}\cdot H^{\frac{2}{n}+1}} \tag{3.1}$$

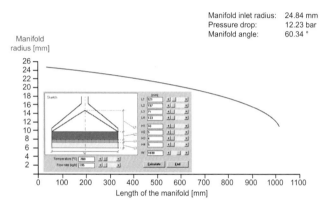

Figure 3.9 Manifold radius as a function of the distance along the length of the manifold

Calculated Example

Calculate Δp for the flat die geometry given in Fig. 3.10 for the conditions:

$\dot{m}$ = 36 kg/h, T = 200 °C, ρ_m = 0.7 g/cm³, H = 2 mm, W = 75 mm.

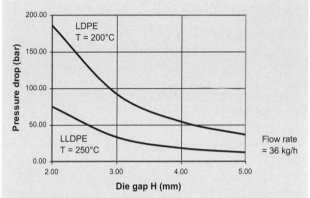

Figure 3.10 Pressure drop Δp as a function of die gap, H, for LDPE and LLDPE, $W = 75$ mm, $L = 100$ mm, $\dot{m} = 36$ kg/h

Viscosity coefficients for LDPE:

$a_0 = 3.388$, $a_1 = -0.635$, $a_{11} = -0.01815$,

$a_2 = -0.005975$, $a_{22} = -0.0000025$, $a_{12} = 0.0005187$

Solution
Volume flow rate

$$\dot{Q} = \dot{m}/\rho_m = 36/(3.6 \cdot 0.7) = 1.429 \cdot 10^{-5} \text{ m}^3/\text{s}$$

Shear rate

$$\dot{\gamma} = 6\dot{Q}/(WH^2) = 6 \cdot 1.429 \cdot 10^{-5}/(0.075 \cdot 0.002^2) = 285.7 \text{ s}^{-1}$$

Power law exponent: $n = 2.75$
Viscosity: $\eta = 646.2$ Pa·s

> Shear stress: $\tau = 184619$ N/m²
>
> Proportionality constant: $K = 8.84 \cdot 10^{-13}$
>
> $\Delta p = 184.62$ bar

Figures 3.10 and 3.11 show Δp calculated according to Eq. 3.1 as a function of the die gap, H, for the given die configuration and the polymers LDPE, LLDPE, and PET.

Flat dies can be easily designed by using the program VISCOAT, listed in Chapter 5. Programmable pocket calculators can handle the calculations as well.

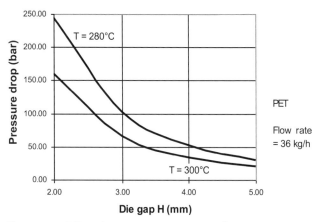

Figure 3.11 Effect of temperature on pressure drop, Δp, $W = 75$ mm, $L = 100$ mm, $\dot{m} = 36$ kg/h

3.5 An Easily Applicable Method of Designing Screen Packs for Extruders

In various extrusion processes, particularly those involving resin blends containing fillers and additives, it is often necessary to increase the melt pressure in order to create more back mixing of the melt in the screw channel of the extruder. This can be achieved by using screen packs of different mesh sizes. They can also be used to increase the melt temperature to attain better plastication of the resin. Another application of screens involves melt filtration, where undesirable material is removed from the melt [29]. In all these operations, it is necessary to be able to predict the pressure drop in the screen packs as accurately as possible, as the melt pressure is closely related to the extruder output. This section presents an easy, quick method of calculating the pressure drop in a screen pack as a function of the resin viscosity, extruder throughput, and the geometry of the screen. The effect of screen blocking is also taken into account. The predictions agree well with the experiments. Examples illustrate the design principles involved.

3.5.1 Mesh Size

The term *screen*, as it is used here, refers to a wire-gauze screen with a certain number of openings depending on the size of the mesh. The mesh size is denoted by the mesh number. This is the number of openings per linear inch, counting from the center of any wire to a point exactly 25.4 mm or one inch distant [29] (Fig. 3.12). The mesh number m_n in Fig. 3.12 for instance, is three. A screen pack usually consists of screens of varied mesh sizes with the screens placed one behind the other in a pack (Fig. 3.13). Figure 3.14 shows the position of the screen pack in an extruder

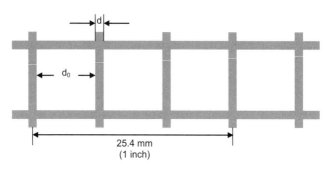

Figure 3.12 Mesh of a wire-gauze screen

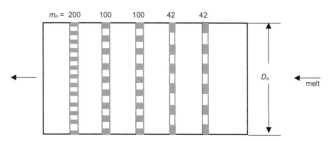

Figure 3.13 Screen pack with screens of varied mesh sizes

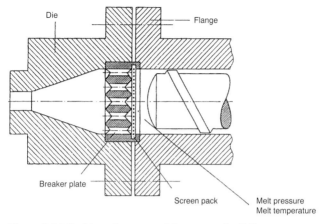

Figure 3.14 Position of screen pack in an extruder [30]

3.5.2 Design Procedure

The volume flow rate $\dot{q}$ through a hole for a square screen opening, is given by [29]

$$\dot{q} = (400 \cdot 6.45 \cdot \dot{M}) / \left(3.6 \cdot \rho_m \cdot m_n^2 \cdot \pi \cdot D_s^2\right) \quad (3.2)$$

The relationship between the volume flow rate $\dot{q}$ and the pressure drop Δp in the screen can be expressed as [10]

$$\dot{q} = K \cdot G^n \cdot \Delta p^n \quad (3.3)$$

The shear rate of the melt flow for a square opening is calculated from [31]

$$\dot{\gamma}_{\text{square}} = (3/0.42) \cdot \dot{q} / \left(0.001 \cdot d_0^3\right) \quad (3.4)$$

The power law exponent n follows from

$$\frac{1}{n} = 1 + a_1 + 2a_{11} \ln \dot{\gamma} + a_{12}T \quad (3.5)$$

where a_1, a_{11}, and a_{12} are viscosity coefficients in the Klein viscosity model, and T is the melt temperature in °F.

Using the Klein model, the melt viscosity η is obtained from

$$\ln \eta = a_0 + a_1 \ln \dot{\gamma} + a_{11} \ln \dot{\gamma}^2 + a_2 T + a_{22}T^2 + a_{12}T \ln \dot{\gamma} \quad (3.6)$$

Shear stress, τ:

$$\tau = \eta \cdot \dot{\gamma} \quad (3.7)$$

Factor of proportionality, K [2]:

$$K = \dot{\gamma}/\tau^n \quad (3.8)$$

The die constant G for a square is calculated from [2]

$$G_{\text{square}} = 0.5 \cdot (0.42/3)^{\frac{1}{n}} \cdot \frac{d_0^{\frac{3}{n}+1}}{2L} \quad (3.9)$$

The length L for a screen is taken as $L = 2d$ with d as the wire diameter (Fig. 3.12). Finally, the pressure drop Δp in the screen results from [2]

$$\Delta p = q^{\frac{1}{n}} \bigg/ \left(K^{\frac{1}{n}} \cdot G_{\text{square}} \right) \quad (3.10)$$

For a screen with openings that has a form other than a square, the formula developed by Schenkel [12] can be applied.

3.5.3 Influence of Polymer Type and Screen Blocking

Calculated Examples

Example 1

For the following input data, it is required to calculate the pressure drop Δp in the screen:

Extruder throughput, $\dot{M} = 454$ kg/h
Melt temperature, $T = 232$ °C
Barrel diameter of the extruder, $D_b = 114.3$ mm
Mesh number of the screen, $m_n = 42$
Melt density, $\rho_m = 0.78$ g/cm³

Solution
From Table 3.1, the minimum sieve opening for a 42 mesh square wire-gauze $d_0 = 0.354$ mm and the wire diameter $d = 0.247$ mm. The screen diameter D_s can be taken as equal to the barrel diameter D_b. Inserting the respective values with the units above, from Eq. 3.2 we get $\dot{q} = 0.00576$ cm³/s and from Eq. 3.4, $\dot{\gamma}_{square} = 928.15$ s⁻¹.

The viscosity η at $\dot{\gamma}_{square} = 928.15$ s⁻¹ is found to be 353.84 Pa·s with $T = 449.6$ °F for the given HDPE.

The power law exponent n follows from Eq. 3.5 to be $n = 1.589$ for the respective resin constants.

Shear stress, $\tau = 328414$ N/m²

Factor of proportionality, K in Eq. 3.8, $K = 1.586 \cdot 10^{-6}$

Die constant G:

$$G_{square} = 0.5 \cdot \left(\frac{0.42}{3}\right)^{\frac{1}{1.589}} \frac{(0.354/100)^{\frac{3}{1.589}+1}}{2 \cdot 2 \cdot (0.247/1000)} = 1.591 \cdot 10^{-8}$$

Finally, the pressure drop Δp in the screen from Eq. 3.10 is $\Delta p = 18.33$ bar.

Example 2

A screen pack consists of two 42-mesh screens, two 100-mesh screens, and one 200-mesh screen (Fig. 3.13). What is the pressure drop in the screen pack for the same conditions as above?

Solution
By applying the equations above, the following values can be obtained for Δp:

for the 42-mesh: 18.3 bar,
for the 100-mesh: 24 bar,
and
for the 200-mesh: 26.9 bar.

The total pressure drop Δp in the screen pack is therefore $\Delta p = 2 \cdot 18.3 + 2 \cdot 24 + 26.9 = 111.5$ bar.

Example 3

What is the pressure drop if the screen is 70% blocked for a screen of mesh size 42 for the same input as above?

Solution
As the diameter of the screen D_s is proportional to the square root of the area, the new diameter for the blocked screen has to be multiplied by the factor $\sqrt{(1-0.7)}$.
This leads to a pressure drop Δp of 24.8 bar for the same conditions as in Example 1.

Result of Calculations

The effect of the type of polymer on the pressure drop in a screen pack consisting of screens of varied mesh size, as shown in Fig. 3.13, is presented in Fig. 3.14. As can be expected, the pressure drop in the screen pack decreases rapidly with decreasing melt viscosity.

Figure 3.15 shows that the influence of the extruder throughput on Δp becomes less pronounced with increasing throughput. The effect of mesh size on Δp is given in Fig. 3.16. The finer gauze with a mesh of 325 has about one and a half times more resistance than a mesh of 42 (see Table 3.2). The exponential increase in pressure with the reduction of screen area, when the screen is blocked by undesirable material, can be seen from Fig. 3.17.

Screen packs are used in extruders for various purposes, such as melt filtration, better back mixing of the melt, and better plastication of the resin. In all these applications, it is important to be able to predict the pressure drop in the screen, as it affects the extruder throughput significantly. The design methods presented here, on the basis of recent advances in computational rheology, accurately simulate the influence of the polymer, mesh size, resin throughput, and screen blocking.

Table 3.2 Dimensions of Square Screens [32]

Mesh size	Sieve opening (mm)	Nominal wire diameter (mm)
42	0.354	0.247
100	0.149	0.110
200	0.074	0.053
325	0.044	0.030

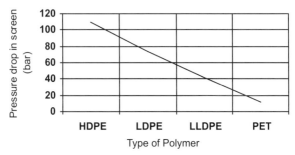

Figure 3.15 Effect of the polymer type on the pressure drop, Δp, in the screen pack

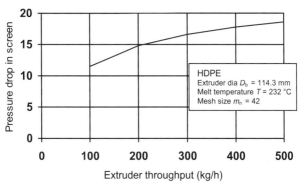

Figure 3.16 Pressure drop in screen vs. extruder throughput

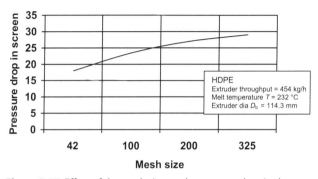

Figure 3.17 Effect of the mesh size on the pressure drop in the screen

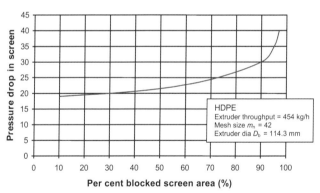

Figure 3.18 Effect of reduced screen area on the pressure drop in the screen

4 Parametrical Studies

Parametrical studies in polymer processing are of importance for determining the effect of significant parameters on the product quality. This section deals with those studies in blown film as an example.

4.1 Blown Film

The blow ratio is the ratio between the diameter of the blown film to the die lip diameter (Fig. 4.1). The effect of operating variables, such as coolant temperature in the case of water-cooled films, on some film properties, such as gloss and haze, is shown in Figs. 4.2 to 4.8 [30]. Depending on the

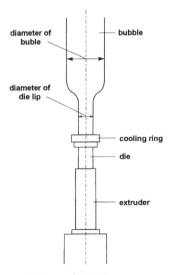

Figure 4.1 Blow-up in blown film [30]

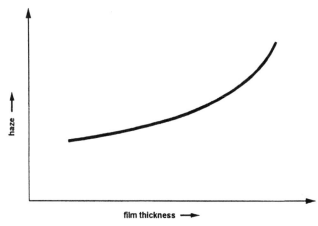

Figure 4.2 Effect of film thickness on haze

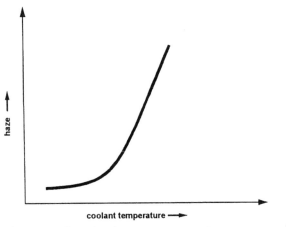

Figure 4.3 Effect of coolant temperature on haze

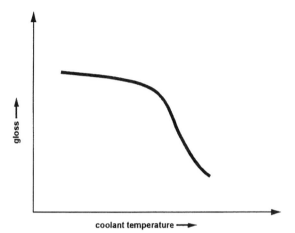

Figure 4.4 Effect of coolant temperature on gloss

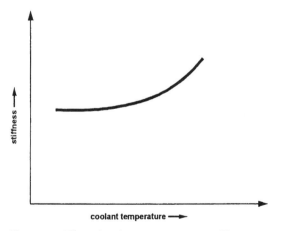

Figure 4.5 Effect of coolant temperature on stiffness

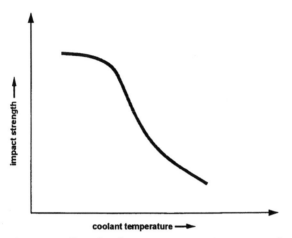

Figure 4.6 Effect of coolant temperature on impact strength

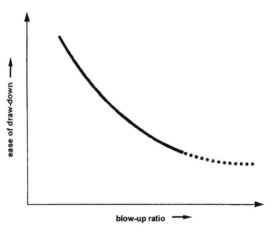

Figure 4.7 Effect of blow-up ratio on ease of draw-down

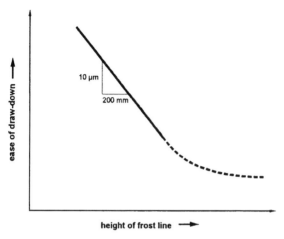

Figure 4.8 Effect of frost line on ease of draw-down

material and the type of film, blow ratios range from 1.3:1 to 6:1 [30].

5 Design Software

The main features of the computer programs mentioned earlier for designing and optimizing extrusion machinery can be summarized as follows:

VISRHEO:
This program calculates the viscosity coefficients occurring in four different viscosity models, namely, Carreau, Muenstedt, Klein, and the power law, which have been treated in Section 1, on the basis of measured flow curves, and stores these coefficients automatically in a resin data bank.

TEMPMELT:
This program calculates the solids melting profile of a multi zone screw taking non-Newtonian melt flow fully into account. The resin data bank containing thermal properties and viscosity constants is part of this program. Melt temperature, pressure, and motor horse power are also calculated.

VISSCALE:
This program is used to scale up multi zone single screws. The resin data bank is included.

VIOUTPUT:
This program determines the output of a single screw extruder for a given screw speed or screw speed for a given output. The resin data bank is part of the program.

VISPIDER:
Program for designing spider dies including resin data bank.

VISPGAP:
Program for designing spiral mandrel dies including resin data bank.

VISCOAT:
Program for designing coat-hanger type dies including resin data bank.

All programs above run on a PC in a user friendly dialogue mode under WINDOWS XP or Vista. The programs can be obtained from Natti S. Rao by contacting his e-mail addresses, raonatti@t-online.de or raonatti77@yahoo.com

5.1 Input and Output Data

In order to describe the functionality of the software, the input and output data of VISRHEO and TEMPMELT are given next in detail. Other programs have already been dealt with in the relevant sections of the book.

First and foremost, it should be mentioned that all programs except VISRHEO, which is a resin databank, have four components, namely, resin selection, geometry of the design element, process parameters and results of calculation.

Consequently, following four simple steps are needed to apply any design program.

STEP 1: Select the resin,
STEP 2: Input the geometry,
STEP 3: Input process parameters, and last,
STEP 4: Calculate and display or print the results.

5.1.1 VISRHEO

As mentioned, VISRHEO evaluates, stores, and retrieves the thermal and rheological properties of any thermoplastic resin.

Thermal Properties

It has been found that the thermal properties vary significantly only with the type of the generic resin, where as the rheological properties are specifically a function the resin brand.

This means that the thermal values are valid for all the brands which fall under a particular generic resin.

However, the rheological data has to be determined experimentally for each resin brand and then input into the databank in the form of coefficients pertaining to the model. The methods of obtaining these coefficients have already been treated in Section 1.3 in Chapter 1.

Figure 5.1 VISRHEO with multiple options of working with this program

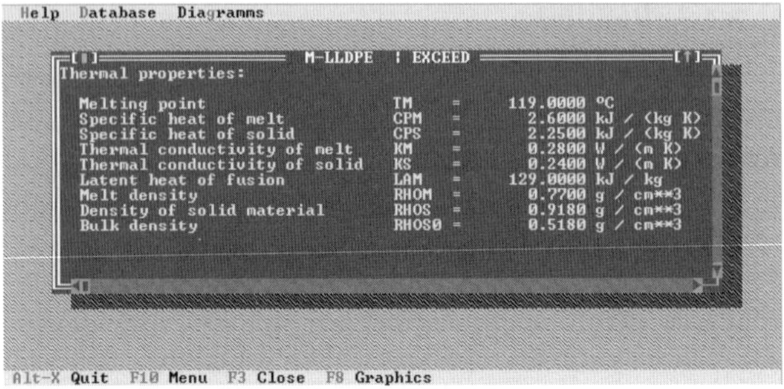

```
Help  Database  Diagramms
┌─[■]─────────────── M-LLDPE  : EXCEED ───────────────[↑]─┐
│ Thermal properties:                                     │
│                                                         │
│   Melting point                      TM    =  119.0000 °C        │
│   Specific heat of melt              CPM   =    2.6000 kJ / (kg K)│
│   Specific heat of solid             CPS   =    2.2500 kJ / (kg K)│
│   Thermal conductivity of melt       KM    =    0.2800 W / (m K) │
│   Thermal conductivity of solid      KS    =    0.2400 W / (m K) │
│   Latent heat of fusion              LAM   =  129.0000 kJ / kg   │
│   Melt density                       RHOM  =    0.7700 g / cm**3 │
│   Density of solid material          RHOS  =    0.9180 g / cm**3 │
│   Bulk density                       RHOS0 =    0.5180 g / cm**3 │
│                                                         │
└─────────────────────────────────────────────────────────┘
Alt-X Quit  F10 Menu  F3 Close  F8 Graphics
```

Figure 5.2 Thermal properties stored in the databank

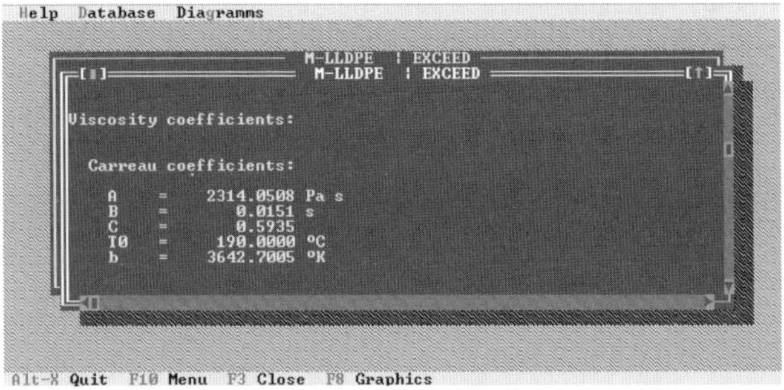

```
Help  Database  Diagramms
                    ┌─────────── M-LLDPE  : EXCEED ───────────┐
┌─[■]──────────────── M-LLDPE  : EXCEED ───────────────[↑]─┐
│                                                          │
│  Viscosity coefficients:                                 │
│                                                          │
│    Carreau coefficients:                                 │
│                                                          │
│       A   =   2314.0508  Pa s                            │
│       B   =      0.0151  s                               │
│       C   =      0.5935                                  │
│       T0  =    190.0000  °C                              │
│       b   =   3642.7005  °K                              │
│                                                          │
└──────────────────────────────────────────────────────────┘
Alt-X Quit  F10 Menu  F3 Close  F8 Graphics
```

Figure 5.3 Sample values of thermal properties

Rheological Data

Examples of rheological input and output data:

From the flow curves of a resin brand (Fig. 1.4), five pairs of shear rate and viscosity are input into the databank at a selected temperature. The program then calculates the constants in the relevant models for that brand, and stores them under the brand name for further design calculations.

5.1.2 TEMPMELT

TEMPMELT is a program for simulating the melting of the resin in an extruder. The dimensions of the suitable screw geometry are derived from the calculated results of the simulations.

Following procedures that are explained by means of figures that show how to input the data, run a simulation, and obtain a display or a printout of the calculated results.

Inputting Screw Geometry

Number of zones:

For a screw as shown in Fig. 2.4, the number is three. If a screw has a shearing and mixing device, as shown in Fig. 2.16, this number equals seven because in this case, the shearing

Figure 5.4 Buttons for performing a screw simulation

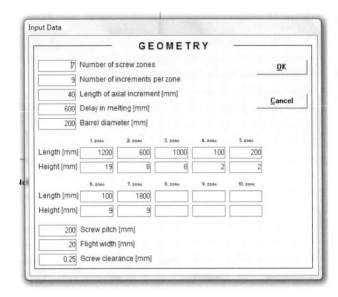

Figure 5.5 Inputting the geometry of a seven-zone screw

element is to be divided into three additional zones in order to provide transition sections to the main shearing area.

Number of increments per zone:

This number is kept at nine.

Length of axial increment:

Input here a value that is one third of the barrel diameter.

Delay in melting: Input here a value that is equal to three times the barrel diameter.

Length: Length of each screw zone

Height: Depth of the channel of each zone at the end of the zone

Screw pitch: For a screw having a square pitch, this value is equal to the screw diameter. For other screws, the corresponding value is to be input.

Flight width: Perpendicular width of the flight (Fig. 2.5)

Screw clearance: Radial clearance between the screw and the barrel.

The screw geometry can also be stored in a file by pushing **Store**, and reloaded into the program by clicking **Load.**

Choose OK when you are done inputting into the Geometry menu and click on the **Polymer** key.

Polymer Resin Selection
Step 1: Select the resin type,
Step 2: Select the resin brand,
Step 3: Select the viscosity model, and
Step 4: Choose **Accept**.

Figure 5.6 illustrates the procedure of selecting the resin.

The program VISRHEO transfers the resin data to TEMPMELT.

Screw speed (rpm):

Output (kg/h): Throughput of the resin

Depending on the button selected, the throughput will be calculated as a function of speed or vice versa. Click on **Update** after the choice has been made. It is also possible to input a desired combination of screw speed and throughput.

Granulate temperature: Self-evident

Barrel temperature: This temperature can be a mean single value or barrel temperature profile.

Temperature tolerance: Value as given in the field.

Initial pressure: Atmospheric pressure

86 Design Software

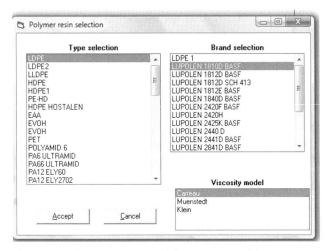

Figure 5.6 Selecting the resin and the viscosity model

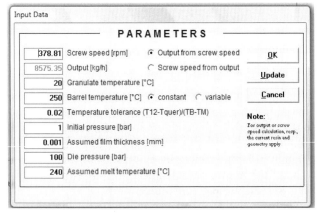

Figure 5.7 Inputting process parameters

Assumed film thickness: This value can vary between 0.001 mm and 0.005 mm, depending on the polymer.

Die pressure: This value depends on the process used.

Assumed melt temperature: The value is equal to the processing temperature of the resin.

After entering all the input data, click **Calculate.**

Results of Simulation

Calculated profiles of the respective screws are given in Figs. 5.8 and 5.10.

How these results can be interpreted has been explained in Section 2.2 on the basis of practical examples.

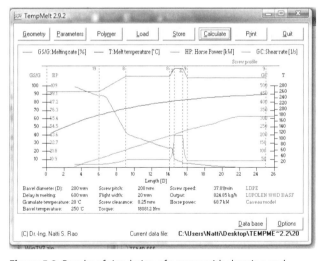

Figure 5.8 Results of simulation of a screw with shearing and mixing devices

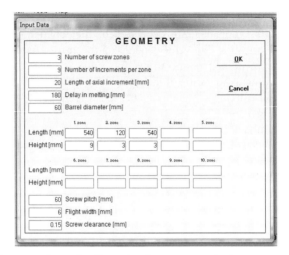

Figure 5.9 Example of inputting the geometry of a three-zone screw

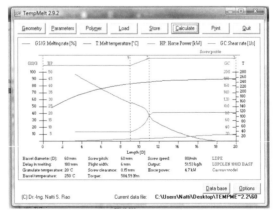

Figure 5.10 Results of simulation of a three-zone screw

6 Thermal Properties of Solid and Molten Polymers

In addition to the mechanical and melt flow properties, thermodynamic data of polymers are necessary for optimizing various heating and cooling processes that occur in plastics processing operations.

In design work the thermal properties are often required as functions of temperature and pressure. As the measured data cannot always be predicted by physical relationships accurately enough, regression equations are used to fit the data for use in design calculations.

6.1 Specific Volume

The volume-temperature relationship as a function of pressure is shown for a semicrystalline PP in Fig. 6.1 [33] and for an amorphous PS in Fig. 6.2. The p-v-T diagrams are needed in many applications, for example to estimate the shrinkage of plastics parts in injection molding [34]. Data on p-v-T relationships for a number of polymers are presented in reference [35].

According to the Spencer-Gilmore equation, which is similar to the van der Waals equation of state for real gases, the relationship between pressure p, specific volume v, and temperature T of a polymer can be written as

$$(v - b^\star)(p + p^\star) = \frac{RT}{W} \tag{6.1}$$

In this equation, $b^\star$ is the specific individual volume of the macromolecule, $p^\star$ the cohesion pressure, W the molecular weight of the monomer, and R the universal gas constant [36].

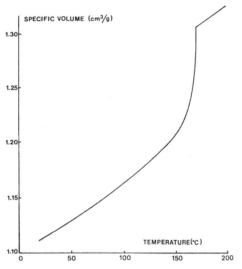

Figure 6.1 Specific volume vs. temperature for a semicrystalline polymer (PP) [33]

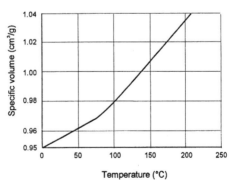

Figure 6.2 Specific volume vs. temperature for an amorphous polymer (PS) [33]

The values p^* and b^* can be determined from p-v-T diagrams by means of regression analysis. Spencer and Gilmore and other workers evaluated these constants from measurements for the polymers listed in Table 6.1 [36, 37].

Table 6.1 Constants for the Equation of State [36]

Material	W g/mol	p^* atm	b^* cm3/g
LDPE	28.1	3240	0.875
PP	41.0	1600	0.620
PS	104	1840	0.822
PC	56.1	3135	0.669
PA610	111	10768	0.9064
PMMA	100	1840	0.822
PET	37.0	4275	0.574
PBT	113.2	2239	0.712

Calculated Example

The following values are given for a PE-LD (LDPE):
$W = 28.1$ g/mol
$b^* = 0.875$ cm³/g
$p^* = 3240$ atm
Calculate the specific volume at
$T = 190\,°C$ and $p = 1$ bar
Using Eq. 6.1 and the conversion factors to obtain the volume v in cm³/g, we obtain

$$v = \frac{10 \cdot 8.314 \cdot (273 + 190)}{28.1 \cdot 3240.99 \cdot 1.013} + 0.875 = 1.292 \text{ cm}^3/\text{g}$$

The density ρ is the reciprocal value of specific volume so that

$$\rho = \frac{1}{v} \tag{6.2}$$

The p-v-T data can also be fitted by a polynomial of the form

$$v = A(0)_v + A(1)_v \cdot p + (2)_v \cdot T + A(3)_v \cdot T \cdot p \tag{6.3}$$

if measured data is available (Fig. 2.3) [9, 38, 39]. The empirical coefficients $A(0)_v \ldots A(3)_v$ can be determined by means of the computer program given in [9]. With the modified two-domain Tait equation [40], a very accurate fit can be obtained both for the solid and melt regions.

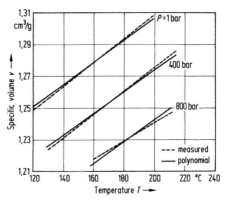

Figure 6.3 Specific volume as a function of temperature and pressure for PE-LD [2, 41]

6.2 Specific Heat

The specific heat c_p is defined as (US uses capital letters instead of h)

$$c_p = \left(\frac{\partial h}{\partial T}\right)_p \tag{6.4}$$

where

h = enthalpy
T = Temperature

The specific heat c_p represents the amount of heat that is supplied to a system in a reversible process at a constant pressure in order to increase the temperature of the substance by dT. The specific heat at constant volume c_v is given by (US uses capital letters instead of u)

$$c_v = \left(\frac{\partial u}{\partial T}\right)_v \tag{6.5}$$

where

u = internal energy
T = Temperature

In the case of c_v the supply of heat to the system occurs at constant volume.

c_p and c_v are related to each other through the Spencer-Gilmore equation, Eq. 6.1:

$$c_v = c_p - \frac{R}{W} \tag{6.6}$$

The numerical values of c_p and c_v differ by roughly 10%, so that for approximate calculations c_v can be made equal to c_p.

Plots of c_p as a function of temperature are shown in Fig. 6.4 for amorphous, semicrystalline, and crystalline polymers.

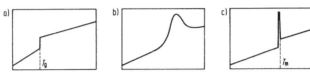

Figure 6.4 (a) Specific heat as a function of temperature for amorphous, (b) semicrystalline, and (c) crystalline polymers [8]

As shown in Fig. 6.5, measured values can be fitted by a polynomial of the type [9]

$$c_p(T) = A(0)\,c_p + A(1)\,c_p \cdot T + A(2)\,c_p \cdot T^2 \tag{6.7}$$

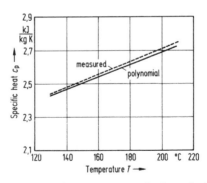

Figure 6.5 Comparison between measured values of c_p [41] and polynomial for LDPE [2]

6.3 Thermal Expansion Coefficient

The expansion coefficient α_v at constant pressure is given by [8]

$$\alpha_v = \frac{1}{v}\left(\frac{\partial v}{\partial T}\right)_p \quad (6.8)$$

The isothermal compression coefficient γ_k is defined as [8]

$$\gamma_k = -\frac{1}{v}\left(\frac{\partial v}{\partial p}\right)_T \quad (6.9)$$

α_v and γ_k are related to each other by the expression [8]

$$c_p = c_v + \frac{T \cdot v \cdot \alpha_v^2}{\gamma_k} \quad (6.10)$$

The linear expansion coefficient α_{lin} is approximately

$$\alpha_{\text{lin}} = \frac{1}{3}\alpha_v \quad (6.11)$$

Data on Coefficients of Thermal Expansion

Table 6.2 shows the linear expansion coefficients of some polymers at 20 °C. The linear expansion coefficient of mild steel lies around $11 \cdot 10^{-6}$ [K^{-1}] and that of aluminum about $25 \cdot 10^{-6}$ [K^{-1}]. As seen in Table 6.2, plastics expand about 3 to 20 times more than metals. Factors affecting thermal expansion are crystallinity, cross-linking, and fillers.

Table 6.2 Coefficients of Linear Thermal Expansion [33, 42]

Polymer	Coefficient of linear expansion at 20 °C α_{lin} 10^6 K^{-1}
PE-LD	250
PE-HD	200
PP	50
PVC-U	75
PVC-P	180
PS	70
ABS	90
PMMA	70
POM	100
PSU	50
PC	65
PET	65
PBT	70
PA6	80
PA66	80
PTFE	100
TPU	150

6.4 Enthalpy

Equation 6.4 leads to

$$dh = c_p \cdot dT \qquad (6.12)$$

As shown in Fig. 6.6, the measured data for $h = h(T)$ [41] for a polymer melt can be fitted by the polynomial

$$h(T) = A(0)_h + A(1)_h \cdot T + A(2)_h \cdot T^2 \qquad (6.13)$$

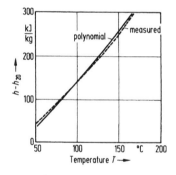

Figure 6.6 Comparison between measured values for h [41] and polynomial for PA6 [9]

The specific enthalpy defined as the total energy supplied to the polymer divided by the throughput of the polymer is a useful parameter for designing extrusion and injection molding equipment such as screws. It provides the theoretical amount of energy required to bring the solid polymer to the process temperature. Values of this parameter for different polymers are given in Fig. 6.7 [8].

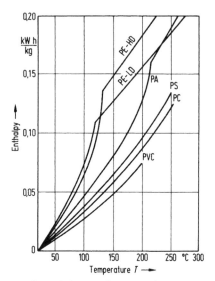

Figure 6.7 Specific enthalpy as a function of temperature [8]

If, for example, the throughput of an extruder is 100 kg/h of polyamide (PA) and the processing temperature is 260 °C, the theoretical power requirement would be 20 kW. This can be assumed to be a safe design value for the motor horse power, although theoretically it includes the power supplied to the polymer by the heater bands of the extruder as well.

6.5 Thermal Conductivity

The thermal conductivity λ (US symbol is k) is defined as

$$\lambda = \frac{Q \cdot l}{t \cdot A \cdot (T_1 - T_2)} \tag{6.14}$$

where

Q = heat flow through the surface of area A in a period of time t

$(T_1 - T_2)$ = temperature difference over the length l.

Analogous to the specific heat c_p and enthalpy h, the thermal conductivity can be expressed as [2]

$$\lambda(T) = A(0)_\lambda + A(1)_\lambda \cdot T + A(2)_\lambda \cdot T^2 \tag{6.15}$$

as shown in Fig. 6.8.

The thermal conductivity increases only slightly with pressure. A pressure increase from 1 bar to 250 bar leads only to an increase in thermal conductivity of less than 5% of its value at 1 bar.

As in the case of other thermal properties, thermal conductivity is, in addition to its dependence on temperature, strongly influenced by the crystallinity and orientation and by the amount and type of filler in the polymer [33]. Foamed plastics have, for example, thermal conductivities at least an order of magnitude lower than those of solid polymers [33].

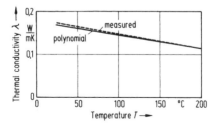

Figure 6.8 Comparison between measured values of λ [41] and polynomial for PP [2]

6.6 Thermal Diffusivity

Thermal diffusivity a (in the US α is used instead of a) is defined as the ratio of thermal conductivity to heat capacity per unit volume [33]

$$a = \frac{\lambda}{\rho \cdot c_p} \qquad (6.16)$$

and is of importance in dealing with transient heat transfer phenomena, such as cooling of melt in an injection mold [2]. Although for approximate calculations average values of thermal diffusivity can be used, more accurate computations require functions of λ, ρ, and c_p against temperature for the solid as well as melt regions of the polymer. Thermal diffusivities of some polymers at 20 °C are listed in Table 6.3 [8, 43].

Exhaustive measured data of the quantities c_p, h, λ, and p-v-T diagrams of polymers are given in the VDMA-Handbook [35]. Approximate values of thermal properties of use to plastics engineers are summarized in Table 6.4 [8, 43].

Experimental techniques of measuring enthalpy, specific heat, melting point, and glass transition temperature by differential thermal analysis (DTA) or differential scanning calorimetry (DSC) are described in detail in [44]. Methods of determining thermal conductivity, p-v-T values, and other thermal properties of plastics are also treated in [44].

Table 6.3 Thermal Diffusivities of Polymers at 20 °C

Polymer	Thermal diffusivity at 20 °C $a\ 10^6\ m^2/s$
PE-LD	0.12
PE-HD	0.22
PP	0.14
PVC-U	0.12
PVC-P	0.14
PS	0.12
PMMA	0.12
POM	0.16
ABS	0.15
PC	0.13
PBT	0.12
PA6	0.14
PA66	0.12
PET	0.11

Thermal Properties of Solid and Molten Polymers

Table 6.4 Approximate values for thermal properties of some polymers [8]

Polymer	Thermal conductivity λ (20 °C) W/m·K	Specific heat c_p (20 °C) kJ/kg K	Density ρ (20 °C) g/cm³	Glass transition temperature T_g °C	Melting point range T_m °C
PS	0.12	1.20	1.06	101	–
PVC	0.16	1.10	1.40	80	–
PMMA	0.20	1.45	1.18	105	–
SAN	0.12	1.40	1.08	115	–
POM	0.25	1.46	1.42	–73	about 175
ABS	0.15	1.40	1.02	115	–
PC	0.23	1.17	1.20	150	–
PE-LD	0.32	2.30	0.92	–120/–90	about 110
PE-LLD	0.40	2.30	0.92	–120/–90	about 125
PE-HD	0.49	2.25	0.95	–120/–90	about 130
PP	0.15	2.40	0.91	–10	160/170
PA6	0.36	1.70	1.13	50	215/225
PA66	0.37	1.80	1.14	55	250/260
PET	0.29	1.55	1.35	70	250/260
PBT	0.21	1.25	1.35	45	about 220

6.7 Coefficient of Heat Penetration

The coefficient of heat penetration is used in calculating the contact temperature that results when two bodies of different temperatures are brought into contact with each other [2, 40].

As shown in Table 6.5, the coefficients of heat penetration of metals are much higher than those of polymer melts. Owing to this, the contact temperature of the wall of an injection mold at the time of injection lies in the vicinity of the mold wall temperature before injection.

Table 6.5 Coefficients of Heat Penetration of Metals and Plastics [39]

Material	Coefficient of heat penetration b $W \cdot s^{0.5} \cdot m^{-2} \cdot K^{-1}$
Beryllium copper (BeCu25)	$17.2 \cdot 10^3$
Unalloyed steel (C45W3)	$13.8 \cdot 10^3$
Chromium steel (X40Cr13)	$11.7 \cdot 10^3$
High density polyethylene (HDPE)	$0.99 \cdot 10^3$
Polystyrene (PS)	$0.57 \cdot 10^3$
Stainless steel	$7.56 \cdot 10^3$
Aluminum	$21.8 \cdot 10^3$

The contact temperature $\theta_{w_{max}}$ of the wall of an injection mold at the time of injection is [39]

$$\theta_{w_{max}} = \frac{b_w \, \theta_{w_{min}} + b_p \, \theta_M}{b_w + b_p} \tag{6.17}$$

where

b = coefficient of heat penetration = $\sqrt{\lambda \rho c}$
$\theta_{w_{min}}$ = temperature before injection
θ_M = melt temperature

The letters w and p refer to mold and polymer, respectively.

Calculated Example

The values given in Table 6.5 refer to the following units of the properties:

Thermal conductivity: W/(m · K)
Density ρ: kg/m³
Specific heat c: kJ/(kg · K)

The approximate values for steel are

l = 50 W/(m · K)
r = 7850 kg/m³
c = 0.485 kJ/(kg · K)

The coefficient of heat penetration is

$$b = \sqrt{\lambda \cdot \rho \cdot c} = \sqrt{50 \cdot 7.85 \cdot 10^3 \cdot 0.485 \cdot 10^3} = 13.8 \cdot 10^3 \; W \; s^{0.5} \cdot m^{-2} \cdot K^{-1}$$

6.8 Heat Deflection Temperature

Heat deflection temperature (HDT) or the deflection temperature under load (DTUL) is a relative measure of a polymer's ability to retain its shape at elevated temperatures for short duration while supporting a load. In amorphous materials the HDT almost coincides with the glass transition temperature T_g. Crystalline polymers may have lower values of HDT but are dimensionally more stable at elevated temperatures [42]. Additives such as fillers have a more significant effect on crystalline polymers than on amorphous polymers. The heat deflection temperatures are listed for some materials in Table 6.6. Owing to the similarity of the measuring principle, Vicat softening point, HDT, and Martens temperature lie often close to each other (Fig. 6.9).

Table 6.6 Heat Deflection Temperatures (HDT) According to the Method A of Measurement [45, 46]

Material	HDT (Method A) °C
PE-LD	35
PE-HD	50
PP	45
PVC	72
PS	84
ABS	100
PC	135
POM	140
PA6	77
PA66	130
PMMA	103
PET	80
PBT	65

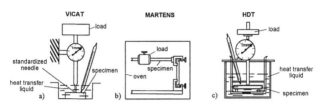

Figure 6.9 Principles of measurement of heat distortion of plastics [45]

6.9 Vicat Softening Point

The Vicat softening point represents the temperature at which a small, lightly loaded, heated test probe penetrates a given distance into a test specimen [42].

The Vicat softening point provides an indication of a material's ability to withstand contact with a heated object for a short duration. It is used as a guide value for the demolding temperature in injection molding. Vicat softening points of crystalline polymers have more significance than amorphous polymers, as the latter tend to creep during the test [42]. Both Vicat and HDT values serve as a basis to judge the resistance of a thermoplastic to distortion at elevated temperatures.

Guide values of Vicat softening temperatures of some polymers according to DIN53460 (Vicat 5 kg) are given in Table 6.7 [45].

The thermodynamic and viscosity data can be stored in a data bank with the help of the software program, VISRHEO, mentioned in Chapter 5 and retrieved for design calculations. Figures 6.10 and 6.11 show samples of the printouts of this program.

Vicat Softening Point

Table 6.7 Guide Values for Vicat Softening Points [45]

Polymer	Vicat softening point °C
PE-HD	65
PP	90
PVC	92
PS	90
ABS	102
PC	138
POM	165
PA6	180
PA66	200
PMMA	85
PET	190
PBT	180

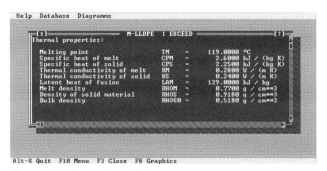

Figure 6.10 Thermodynamic data

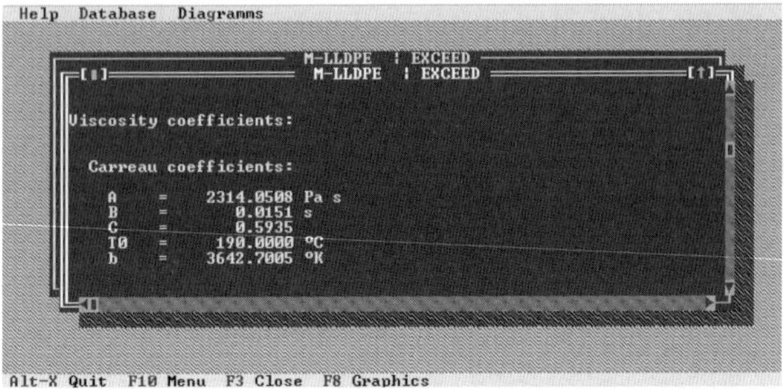

Figure 6.11 Viscosity coefficients in the rheological models

7 Heat Transfer in Plastics Processing

Heat transfer and flow processes occur in most polymer processing machinery and often determine the production rate. Designing and optimizing machine elements and processes therefore requires the knowledge of the fundamentals of these sciences. The flow behavior of polymer melts has been dealt with in Chapter 1. In the present chapter, the principles of heat transfer of relevance to polymer processing are treated with examples.

7.1 Case Study: Analyzing Air Gap Dynamics in Extrusion Coating by Means of Dimensional Analysis

A number of experimental investigations in the past have dealt with the influence of various parameters, such as melt temperature, air gap distance, and coating thickness on the bonding strength between a primed film and the coating. This section presents a general procedure for obtaining a quantitative relationship, in order to predict the influence of individual parameters on the bond strength on the basis of dimensional analysis. The applicability of this method was demonstrated by evaluating the experimental results published in the literature, which led to a practical formula for predicting adhesion as a function of the parameters mentioned.

It was found that for a given temperature the shear history and the residence time of the melt in the air gap are of utmost importance to the adhesion between the film and the coating. The application of the formulas given is explained by a number of numerical examples.

Laminating and coating processes involve a polymer film moving from a flat die at a high temperature and speed on to the rolls (Fig. 7.1). Experiments show that, once the film is successfully primed, the bond quality between the film and the coating depends on factors such as melt temperature, film thickness, and oxidation in the air gap, to mention a few [47].

In this section, first the heat transfer between the film and the surrounding air is mathematically described, in order to predict the cooling of the film in the air gap. It was then shown that the bond strength depends on the shear history of the melt and the residence time of the film in the air gap for a given temperature by analyzing the experimental results on the basis of dimensional analysis.

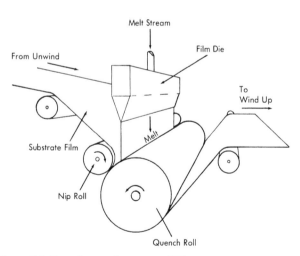

Figure 7.1 Extrusion coating process [48]

7.1.1 Heat Transfer between the Film and the Surrounding Air

The unsteady-state heat transfer between the film and the air by heat conduction can be calculated by means of the overall heat transfer coefficient [49] given by

$$k = \alpha_i = 6\,\lambda/s \qquad (7.1)$$

k = overall heat transfer coefficient (W/m² K)
α_i = internal heat transfer coefficient (W/m² K)
λ = thermal conductivity (W/m · K)
s = film thickness (m)

Substituting relevant values for a LDPE film with a thickness of 20 µm, we get from Eq. 7.1 with $\lambda = 0.24$ W/m · K

$$\alpha_i = 6 \cdot 0.24 / 20 \cdot 10^{-6} = 72\,000 \text{ W/m}^2 \cdot \text{K}$$

However, when the film moves in air, the external heat transfer coefficient α_a will be only of the order of 10 W/m² K [40] and determines the heat transfer between the film and the air. In other words, the heat from the film is controlled by the external resistance.

For this case, the temperature of the film in the air gap can be calculated from

$$\frac{T - T_\infty}{T_A - T_\infty} = \exp\left(-\frac{k \cdot A}{\rho \cdot c_p \cdot V} \cdot t\right) \qquad (7.2)$$

where

T = final film temperature
T_∞ = air temperature

T_A = initial temperature of the film (melt temperature)
t = time in the air gap

Calculated Example

$T_A = 200\ °C$, $T_\infty = 20\ °C$, $k = \alpha_a = 10\ W/m^2\ K$, $\rho = 700\ kg/m^3$, $c_p = 2.3\ kJ/kg\ K$, and $t = 200$ ms. Calculate the final film temperature.

Solution
The ratio of the volume V to the heat transfer area A can be substituted by [49]

$$V/A = 2\ s \tag{7.3}$$

where s = film thickness.

The film thickness can be obtained from the coat weight cwt

$$s = 1000\ \text{cwt}/\rho \tag{7.4}$$

cwt = coat weight (g/m²)
ρ = density (kg/m³)

For a coat weight of 10 g/m², the film thickness follows from Eq. 7.1 with $\rho = 700\ kg/m^3$ for LDPE melt $s = 14.28\ \mu m$. Using a film thickness $s = 30\ \mu$ and $k = 10\ W/m^2\ K$, the dimensionless expression $\dfrac{k \cdot A/V}{\rho \cdot c_p}$, called number of transfer of units, equals 0.02.

The right hand side of Eq. 7.1 equals 0.98, which leads to a final temperature of 196.4 °C for the conditions in the example. Thus, the cooling of the film in the air gap is not significant. However, at thicker films it can be considerable, and can affect the chemical kinetics of adhesion. In the evaluation of the experiments treated below, the final film temperature is assumed to be equal to the melt temperature.

7.1.2 Chemical Kinetics

The effect of chemical kinetics on the adhesion was taken into account by means of the shift factor a_T [50, 51]

$$a_T = b_1 \cdot \exp\left[b_2 / (T + 273)\right] \tag{7.5}$$

Using the constants $b_1 = 5.13 \cdot 10^{-6}$ and $b_2 = 5640$ K, the shift factors at various melt temperatures were calculated.

Calculated Example

> The shift factor at the melt temperature of $T = 321\,°C$ amounts to
>
> $a_T = 5.13 \cdot 10^{-6} \cdot \exp(5640 / 594) = 0.068$

Time in the Air Gap (TIAG)

The Graetz number is defined as the ratio of the time to reach thermal equilibrium perpendicular to the flow direction to the residence time. The time in the air gap can be taken as the residence time and its effect on the bond strength predicted by means of the Graetz number Gz. This number can be calculated from [52]

$$Gz = \dot{m} \cdot c_p / \lambda \cdot GAPL \tag{7.6}$$

with

$\dot{m}$ = throughput (kg/h)
c_p = specific heat (kJ/kg K)
λ = thermal conductivity (W/m K)
$GAPL$ = Length of air gap (m)

Calculated Example

Calculate Gz for m = 197 kg/h, c_p = 2.2 kJ/kg K, λ = 2.2 W/m K, and $GAPL$ = 102 mm. Using coherent dimensions Gz follows from Eq. 7.6:

$$Gz = 10 \cdot 197 \cdot 2.2 / 36 \cdot 0.24 \cdot 0.102 = 4918$$

Shear History of the Film

It was found [47] that the shear history of the film plays a significant role in the gap dynamics. The shear can be calculated by multiplying the shear rate of the film at the die exit and time in the air gap (TIAG).

$$\text{SHEAR} = \text{Shear rate at the die exit} \cdot \text{TIAG}$$

7.1.3 Evaluation of the Experiments

The experimental results of Ristey and Schroff [53] were evaluated using the concepts described above. The Graetz numbers calculated from the experiments lie in the range of 1600 to 5000. A shear rate range of 69 s^{-1} to 208 s^{-1} was taken as a basis.

By applying stepwise linear regression, the following equation has been developed from the experimental results

$$\ln(C = 0) = -6.7684 - 3.1784 \ln(\text{shift factor}) \quad (7.7)$$
$$- 0.8487 \ln(Gz) + 1.2784 \ln(\text{shear})$$

with $C = 0$: Carbonyl Absorbance

Analyzing Air Gap Dynamics in Extrusion Coating

The 45 degree slope in Fig. 7.2 shows the calculated and experimental results. Owing to the complicated processes involved in the air gap dynamics, there is some deviation between calculations and measurements.

Still, it is seen that Eq. 7.7 predicts the combined effect of the significant parameters not only in the trend but in quantitative terms as well close to reality.

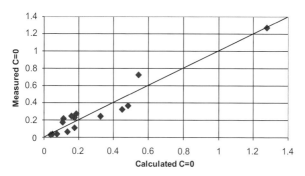

Figure 7.2 Comparison between measured [52] and calculated carbonyl absorbance

Further Reading

Rao, N. S., Schott, N. R., Understanding Plastics Engineering Calculations (2012) Hanser Publishers, Munich

Rao, N. S., O'Brien, K. T., Design Data for Plastics Engineers (1998) Hanser Publishers, Munich

References

[1] Laun, H. M., *Prog. Colloid Polym. Sci.* (1987) 75, p. 111
[2] Rao, N., Design Formulas for Plastics Engineers (1991) Hanser, Munich
[3] Muenstedt, H., *Kunststoffe* (1978) 68, p. 92
[4] Ostwald, W., *Kolloid-Z.* (1925) 36, p. 99
[5] de Waele, A., *J. Oil Colour Chem. Assoc.* (1923) 6, p. 33
[6] Klein, I., Marshall, D. I., Friehe, C., *SPE J.* (1965) 21, p. 1299
[7] N. N., Brochure, Kunststoff-Physik im Gespraech, BASF (1977)
[8] Rauwendaal, C., Polymer Extrusion, 4th ed. (2001) Hanser, Munich
[9] Rao, N. S., Designing Machines and Dies for Polymer Processing (1981) Hanser, Munich
[10] Procter, B., *SPE J.* (1972) 28, p. 34
[11] Bernhardt, E. C., Processing of Thermoplastic Materials (1963) Rheinhold, New York
[12] Schenkel, G., Kunststoff-Extrudiertechnik (1963) Hanser, Munich
[13] Rao, N. S., Ruegg, T., *TAPPI PLC Conf.* (2006) Boston
[14] N. N., Brochure, Eastman (1997)
[15] Tadmor, Z., Klein, I., Engineering Principles of Plasticating Extrusion (1963) Van Nostrand Rheinhold, New York
[16] Fritz, H. G., in Extrusion Blow Molding in Plastics Extrusion Technology, Hensen, F. (Ed.) (1988) Hanser, Munich
[17] Darnell, W. H., Mol, E. A., *SPE J.* (1956) 12, p. 20
[18] Pearson, J. R. A., Reports of University of Cambridge, Polymer Processing Research Centre (1969)

[19] Fischer, P., Dissertation, RWTH Aachen (1976)
[20] Potente, H., Proceedings of 9th Kolloquium IKV, (1976) Aachen
[21] Rao, N., Hagen, K., Kraemer, A., *Kunststoffe* (1979) 69, p. 173
[22] Rao, N. S., O'Brien, K. T., Harry, D. H., in Computer Modeling for Extrusion and Other Continuous Polymer Processes, O'Brien K. T. (Ed.) (1992) Hanser, Munich
[23] N. N., Brochure, BASF (1982)
[24] Ast, W., *Kunststoffe* (1976) 66, p. 186
[25] Sammler, R. L., Koopmans, R. J., Magnus, M. A., Bosnyak, C. P., *ANTEC, Conf. Proc.* (1998) p. 957
[26] Rosenbaum, E. E. et al., *ANTEC, Conf. Proc.* (1998) p. 952
[27] N. N., Brochure, Blow Molding, BASF (1992)
[28] Chung, C. I., Lohkamp, D. T., *SPE 33*, ANTEC 21 (1975) p. 363
[29] Carley, J. F., Smith, W. C., *Polym. Eng. Sci.*, (1978) 18, p. 408
[30] N. N., Brochure, BASF (1992)
[31] Ramsteiner, F., *Kunststoffe* (1971) 61, p. 943
[32] Perry, R. H., Green, D. W., Perry's Chemical Engineers' Handbook (1984) McGraw-Hill
[33] Birley, A. W., Haworth, B., Bachelor, T., Physics of Plastics (1991) Hanser, Munich
[34] Kurfess, W., *Kunststoffe* (1971) 61, p. 421
[35] N. N., Kenndaten für die Verarbeitung thermoplastischer Kunststoffe: Thermodynamik (1979) Hanser, Munich
[36] McKelvey, J. M., Polymer Processing (1962) John Wiley, New York
[37] Progelhof, R. C., Throne, J. L., Polymer Engineering Principles – Properties, Processes, Tests for Design (1993) Hanser, Munich
[38] Münstedt, H., Berechnen von Extrudierwerkzeugen (1978) VDI, Düsseldorf
[39] Wübken, G., Berechnen von Spritzgießwerkzeugen (1974) VDI, Düsseldorf
[40] Martin, H., VDI-Wärmeatlas (1984) VDI, Düsseldorf
[41] N. N., *Kunststofftech. Kolloq. IKV Aachen, [Ber.], 9th* (1978)
[42] N. N., General Electric Plastics Brochure, Engineering Materials Design Guide

References

[43] Ogorkiewicz, R. M., Thermoplastics Properties and Design (1973) John Wiley, New York
[44] N. N., Brochure, Advanced CAE Technology Inc. (1992)
[45] Handbuch der Kunststoffprüfung, Schmiedel, H. (Ed.) (1992) Hanser, Munich
[46] Domininghaus, H., Plastics for Engineers (1993) Hanser, Munich
[47] Bender, E., Lecture notes, Wärme and Stoffübergang, Univ. Kaiserslautern (1982)
[48] Osborne, K. R., Jenkins, W. A., Plastic Films (1992) Technomic, Lancaster, PA
[49] McCabe, W. L., Smith, J. C., Harriott, P., Unit Operations of Chemical Engineering (1985) McGraw-Hill, New York
[50] Welty, J. R., Wicks, C. E., Wilson, R. E., Fundamentals of Momentum, Heat and Mass Transfer (1983) John Wiley, New York
[51] Kreith, F., Black, W. Z., Basic Heat Transfer (1980) Harper & Row, New York
[52] Thorne, J. L., Plastics Process Engineering (1979) Marcel Dekker Inc., New York
[53] Ristey, W. J., Schroff, R. N., *TAPPI Pap. Synth. Conf., [Proc.]* (1978), p. 267

Index

A

air gap 109
annulus 23, 25, 53
apparent shear rate 12
apparent viscosity 13

B

barrel temperature 31
blow molding dies 56
blown film 73
blow ratio 73
bond strength 109

C

circle, channel shape 23
coating 45, 109
coating processes 110
coefficient of heat penetration 104
contact temperature 103
coolant 73

D

deflection temperature under load (DTUL) 104
demolding temperature 106

die constant 21, 66
die geometry 56

E

enthalpy 93, 97, 99, 100
expansion coefficient 95
extrusion 27, 28, 97
extrusion coating 59
extrusion dies 22, 49
extrusion screw 39, 44

F

feed zone 29, 30
flat dies 59, 62
flow processes 109
flow rate 12, 19, 20

G

geometry constant 23
Graetz number 113, 114

H

haze 73
heat deflection temperature (HDT) 104

Index

heat penetration 103
heat transfer 109, 111
heat transfer coefficient 111

I

internal energy 93
isothermal compression coefficient 95

K

Klein viscosity model 16, 66

L

laminar flow rate 21
laminating 110
linear expansion coefficient 95

M

manifold radius 59
Martens temperature 105
mass flow rate 19
melt flow 19
melt flow index 19
melt flow rate 19
melt fracture 56
melting profile 39, 79
melt temperature 104

melt viscosity 11, 69
melt volume index 19
melt volume rate 19
mesh size 63

P

pelletizer dies 56
power law 14
power law exponent 58
pressure 49
pressure drop 7, 21, 26, 50, 54, 60, 63
production rate 109

R

radius 12, 25
residence time 49, 109, 110, 113
rheological data 82

S

screen 63
shear history 114
shear rate 34, 49, 61, 82, 114
slit 23
specific volume 89
spider dies 49, 79
spiral mandrel dies 53
square, channel shape 23

T

TEMPMELT 79, 83
thermal conductivity 98, 111
thermal diffusivity 100
thermal expansion 95
thermal properties 81
three-zone screw 28
triangle, channel shape 23

V

Vicat softening point 105, 106
VIOUTPUT 79
VISCOAT 80
viscosity 82
viscosity coefficient 66
VISPGAP 80
VISPIDER 79
VISRHEO 79, 81
VISSCALE 79
volumetric flow rate 21

HANSER

Hands-on Examples and Case Studies.

Rao/Schott
Understanding Plastics Engineering Calculations
206 pages
ISBN 978-1-56990-509-8

Starting from practical design formulas which are easily applicable, and yet take the resin rheology into account, this guide provides answers to these questions quickly and effectively by guiding the user step by step through the computational procedures on the basis of illustrative technical examples.
All the calculations involved can be handled by pocket calculators and hence can be performed right on the site where the machines are running. This guide is a valuable tool not only to troubleshoot but also to estimate the effect of design and process parameters on the product quality in plastics processing.

More Information on Plastics Books and Magazines:
www.hanserpublications.com or **www.kunststoffe-international.com**

HANSER

Always the right Formula.

Rao/Schumacher
**Design Formulas
for Plastics Engineers**
2nd edition
176 pages
ISBN 978-1-56990-370-4

This book presents a summary of the most important formulas and their applications to solve design and processing problems with plastics materials. Numerous practical examples guide the reader step-by-step through the computational routine of designing polymer machinery. The approach does not assume prior knowledge and is very practical to enable every engineer to apply these concepts in their daily work, improve their equipment, and stabilize their processes.

More Information on Plastics Books and Magazines:
www.hanserpublications.com or **www.kunststoffe-international.com**

HANSER

Extrusion 101

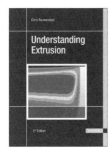

Rauwendaal
Understanding Extrusion
2nd edition
248 pages. Four-color
ISBN 978-1-56990-453-4

This book presents basic information on extrusion technology and is accessible to professionals without an engineering degree. Written for extruder operators, supervisors, and technical service professionals, but also for newcomers to the industry and students, it introduces the process, the machinery, and general information on process control, materials, and troubleshooting.
The second edition is extended to cover high-speed extrusion, how to reduce material cost, efficient extrusion, purging and product changeover, how to reduce energy consumption, new developments in extruder screw design, and more.

More Information on Plastics Books and Magazines:
www.hanserpublications.com or www.kunststoffe-international.com

HANSER

No More Problems with Single-Screw Extruders

Campbell/Spalding
Analyzing and Troubleshooting Single-Screw Extruders
800 pages
ISBN 978-1-56990-448-0

Prior extrusion books are based on barrel rotation physics – this is the first book that focuses on the actual physics of the process-screw rotation. In the first nine chapters, theories and math models are developed. Then, these models are used to solve actual commercial problems in the remainder of the book. Realistic case studies are presented that are unique in that they describe the problem as viewed by a typical plant engineer and provide the actual dimensions of the screws. The new knowledge in this book will be highly useful for production engineers, technical service engineers, consultants specializing in troubleshooting and process design, and process researchers and designers.

More Information on Plastics Books and Magazines:
www.hanserpublications.com or **www.kunststoffe-international.com**

HANSER

A Comprehensive Account of the Full Range of Dies Used for Extrusion.

Michaeli
**Extrusion Dies
for Plastics and Rubber**
3rd edition
362 pages
ISBN 978-1-56990-349-0

In this well-received book the distinctive features of the various types of dies are described in detail. Advice on the configuration of dies is given, and the possibilities of computer-aided design, as well as its limitations, are demonstrated. The fundamentals and computational procedures are well explained so that the reader does not need any special prior knowledge of the subject. The mechanical configuration, handling, and maintenance of extrusion dies are described. Calibration procedures for pipes and profiles are also dealt with. This book was written for plastics engineers as well for students preparing for their professional life.

More Information on Plastics Books and Magazines:
www.hanserpublications.com or **www.kunststoffe-international.com**